Home Generator Selection, Installation, and Repair

Home Generator Selection, Installation, and Repair

Paul Dempsey

New York   Chicago   San Francisco   Athens   London   Madrid
Mexico City   Milan   New Delhi   Singapore   Sydney   Toronto

Home Generator Selection, Installation, and Repair

1 2 3 4 5 6 7 8 9 0   DOC/DOC   1 9 8 7 6 5 4 3

ISBN   978-0-07-181297-9
MHID   0-07-181297-0

This book is printed on acid-free paper.

Sponsoring Editor   **Copy Editor**
Judy Bass   James Madru

Editing Supervisor   **Proofreader**
Stephen M. Smith   Claire Splan

Production Supervisor   **Indexer**
Richard C. Ruzycka   Judy Davis

Acquisitions Coordinator   **Art Director, Cover**
Amy Stonebraker   Jeff Weeks

Project Manager   **Composition**
Patricia Wallenburg, TypeWriting   TypeWriting

About the author

Paul Dempsey is a master mechanic and a former magazine editor. He is the author of some 30 technical books with subjects ranging from bicycles to heavy-duty diesel engines. He is the owner of a shop, located in a rural area south of Veracruz, Mexico, that aspires to be a universal repair station. Almost anything that comes in the door—generators, tractors, marine engines, garden equipment—can be fixed. The shop's collection of antique machine tools enables many parts unobtainable in Mexico to be fabricated.

Contents

3 • Generator repairs   *37*

4 • Engine electrical   77

Preface

Millions of Americans depend on portable generators to provide power during the blackouts that have become increasingly frequent. According to the Eaton Corporation, the number of outages affecting 50,000 people or more doubled between 2010 and 2011. A half-million people lose grid power every day in this country.

There are various reasons why the North American grid, once the envy of the world, can no longer keep pace with demand. Some critics blame poor maintenance, others see privatization—80 percent of the grid is privately owned—as the culprit, and still others point to the high level of integration that can escalate local failures into regional catastrophes. An example of the latter occurred in August 2003. A high-voltage transmission line sagged, touched a wet tree branch, and shorted to ground. This produced a cascade of exploded transformers and burned-out wiring that shut off power to 50 million people in eight states and parts of southern Canada. The outage cost several billion dollars and the lives of 11 people.

A portable generator—one with good build quality, sized for the loads, and properly maintained—will keep the lights on. But generating your own electricity is not a simple, "plug in and play" proposition. This book describes:

- *How to purchase the right generator for your needs.* The first requirement is to make sense of the esoteric engineering terms used in the sales literature. Next, one must determine how much electricity is needed during a blackout. Do you want to power your whole house? If not, which loads are critical? The more power required, the greater the cost.

  Most people specify a gasoline-powered generator, but diesel, propane, and natural gas have advantages for certain applications. The best diesel generators are industrial products that should give decades of reliable service.

Then there is the question of quality. Many generators are poorly constructed devices without much of anything by way of factory support. The book describes ways to evaluate build quality before purchase.

Certain features, such as large, easy-rolling, pneumatic tires; steel—not plastic—fuel tanks; basic instrumentation; and ground fault protection on output circuits, are important and worth paying extra for.

- *How to connect the generator.* There are various ways, each with advantages and disadvantages, to safely connect auxiliary power to your home or office. An improperly connected generator can impose serious risks, both to yourself and to linemen working to restore grid power. You will need to purchase an appropriate power cord and make provision for storing fuel, and you should have some spare parts, such as filters and spark plugs, on hand.
- *How to troubleshoot and repair generator malfunctions.* For the most part, portable generators are simple machines, quite within the capabilities of do-it-yourself mechanics to repair. The book goes into detail on how generators work and how to recognize and test various circuit components. It also includes a comprehensive list of parts suppliers.
- *How to troubleshoot and repair generator engines.* The engines that power generators spend much of the time idle, waiting for a blackout. When the lights go down, they work hard at something like 80 or 90 percent of maximum power. The combination of long periods of storage followed by near full-throttle operation is a recipe for trouble. Unless you buy a state-of-the-art industrial generator, you will sooner or later be working on the engine.

This section of the book describes how to identify the causes of gasoline and diesel engine malfunctions and provides detailed repair information. The material applies to all small engines, not merely to those that power generators.

The book is written by a DIY mechanic for other DIY mechanics who do not have access to factory tools or training. I have tried to illustrate all critical parts, so that nonprofessionals can recognize what they're dealing with. In so far as space permits, wiring diagrams are provided. Special tools are described in enough detail to enable readers to fabricate their own.

Overall, the emphasis is on safety. Portable generators present hazards, both when using them and, especially, during repair operations, some of which must be accomplished while the generator is running.

Paul Dempsey

1

Selecting the right generator

In September 1882, a 100-kW generator came on line to provide electric power to a few square blocks in lower Manhattan. Factories, hotels, and a few ships previously had been electrified by their owners. But Thomas Edison's Pearl Street Station was the first to generate power for anyone within range who was willing to pay for it.

Pearl Street turned a profit within two years, and other entrepreneurs set up local generating stations, most of which used alternating current (AC) rather than the 110-V direct current (DC) favored by the inventor. Alternating current is variable-voltage power that is easy to step up for transmission and to step down for safe residential use By the 1930s, the local generating stations were combined into regional distribution networks that would coalesce a few decades later into the three main grids that serve the continental United States and parts of Canada and northern Mexico (Fig. 1-1).

Power for the grid comes from more than 12,000 generating units:

- Base-load units operate continuously to meet normal demand.
- Peaking generators are dispatched during periods of high demand, such as late-afternoon summer days.
- Intermediate, or cycling, units bridge the gap between normal and peak demand.

Because coal is cheap, most base-load generators are coal-fired (Fig. 1-2). Peaking generators, consisting for the most part of oil-fueled turbines, are the most expensive to run. In the past, natural gas was pretty well confined to intermediate units, but recent declines in natural gas prices make the fuel

1

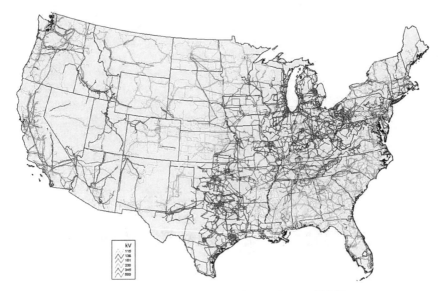

FIG. 1-1 *The national grid reflects past population density.* FEMA

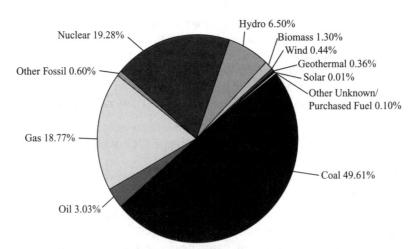

FIG. 1-2 *Coal remains the dominant generator fuel, used in nearly all baseline plants.* EPA

practical for base-load generation. Although wind and solar generators have the lowest operating costs, utilities have difficulty integrating these variable-output devices into their systems. The wind blows as it will, and the sun goes down every night.

Outages

Unfortunately, the North American grid, considered by many to be the world's largest and most complex machine, has not aged well. One indication is the number of power interruptions. According to the Eaton Corporation, the frequency of blackouts affecting more than 50,000 people nearly doubled between 2010 and 2011. And repairs are slow. Three Carnegie Mellon professors found that power outages last seven times longer in the United States than in the Netherlands and four times longer than in France. The study was done in 2006; since then, response time has slowed. The American Society of Civil Engineers gives the grid a grade of D+. Should you need any more convincing, the World Economic Forum puts the American grid in thirtieth place, behind the distribution systems of 29 other countries, some of them barely industrialized.

Most of us are all too familiar with blackouts. For those who aren't, it's worth noting that when the power goes down, so does nearly everything else. Urban residents experience a loss of water pressure, and rural residents who depend on wells do without water. If a blackout occurs in summer, air conditioners and fans go out, which means that office buildings and most modern homes become uninhabitable. Heat stroke is a real danger when the humidity is high and ambient temperatures are above body heat. In winter, central heating systems no longer function. People take refuge in their automobiles, running the heaters so long as they have gasoline.

Nor can one easily escape an area-wide blackout. Freeways clog, traffic signals no longer function, and gas stations cannot pump fuel from their underground tanks. Local radio and television stations go off the air, and cell phones may not work. A blackout of any length represents a return to the past, but without the amenities, the ways of coping, that people in earlier centuries enjoyed.

Most short-term outages come about because of routine weather events— high summer temperatures result in greater demands for electricity than overworked transmission systems can deliver or a cold snap ices over and breaks power lines. Systems also go down during the first rain after a long dry spell. The accumulated dust becomes conductive, and transformers short out, often explosively.

Intrusions are another source of brief blackouts. Birds get themselves across power lines, a truck knocks down a transmission tower, or a backhoe operator cuts an underground line. Intrusions are impossible to prevent, but happen frequently enough to be predictable. For example, the Edison Electric

Institute blames foraging animals for 11 percent of U.S. blackouts. Most of us cope with these short-term inconveniences.

The triggers for long-term catastrophic outages usually take the form of hurricanes, tornados, forest fires, and violent wind storms (Fig. 1-3). These events occur more frequently as the climate grows warmer. We see droughts in the Southwest; massive, uncontrollable forest fires in the West; and Noah-like floods in other parts of the country. The recent *derecho* put life on hold for millions of people in 10 states and the District of Columbia, where politicians, some of them anyway, still manage to deny that something has gone horribly askew with the climate. But people charged with keeping the power on do not have the luxury of denial. As Bill Gausman, a senior vice president at Potomac Electric, told *The New York Times*, "We've got a 'hundred-year storm' every year now." A hundred-year-storm is the engineering standard of safety for buildings, bridges, offshore drilling platforms, and electrical transmission systems. Hurricane Sandy, which flooded New York City subways for the first time in their 104-year history and left more than 8 million people without power, was a confirmation of Mr. Gausman's statement.

Even if the climate had not turned pathologic, utilities would be in trouble. Privatization, a movement that began in the 1970s and continues to accelerate, makes investor profits central. From 1975 to 2000 (the last year for which we have data), utilities spent $117 million less on equipment upgrades than during the year before. As a result, the average substation transformer is two years past its design life, and three-quarters of the high-voltage transmission

FIG. 1-3 *Hurricane Sandy photographed as it neared the East Coast on October 29, 2011. By the next day, 8.5 million people were without power, and some would wait weeks for power to return.* NOAA

transformers are more than 25 years old. The lead time—the time from when an order is placed to delivery—for these massive 800,000-lb transformers is two years or more.

Meanwhile, demand increases at the rate of more than 2 percent a year, complicated by the population shift to the South and Southwest, where air conditioning makes life tolerable (Fig. 1-4). Base-load power plants, the structural core of the industry, require a decade or more to build. Nuclear plants, which currently produce about 20 percent of our base-line power, cost $7000 per kilowatt (1 kW, or kilowatt = 1000 W). Coal-fired plants cost the same with the necessary emission controls.

Add to this the pending legislation to limit airborne carbon releases and the uncertainties involved with decommissioning nuclear power plants, many of which are approaching the end of their useful lives, and we can understand why utilities and rural co-ops encourage conservation and on-site power generation. Even so, the appeal to conservation sounds strange: it is as if General Motors were to tell its customers to start walking or to build their own cars.

In a sense, we're back where we started more than 120 years ago. Individuals now must make their own provisions for electric power. As a consequence, residential generator sales are booming. Three percent of upscale homes now come with standby power. Hundreds of imported generators flood the market, and Generac—by most measures the leading American

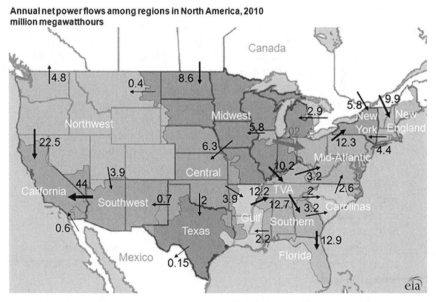

Annual net power flows among regions in North America, 2010
million megawatthours

FIG. 1-4 *The population shift to the South and Southwest puts customers farther from generating plants and overloads existing transmission lines.* U.S. Energy Information Administration

manufacturer—operates its three Wisconsin plants around the clock to meet the demand.

What's available

Portable generators, or *gensets* (the terms are synonymous), range in capacity from 0.5 to more than 17 kW. The smaller units can be carried by hand; larger models have wheels, either as a standard feature or as an extra-cost option. The largest portables can power an entire home almost as if the grid were still functioning. But most people find that they can get by with 6 kW or less. A limited amount of electricity is infinitely better than none.

Some of these machines would do Thomas Edison proud; others are, to put it plainly, junk that would be hard-pressed to survive a week-long blackout.

In terms of operation, gensets fall into three groups:

- *Brushed generators* represent the older technology. Edison's Pearl Street generator used the same carbon brushes as these machines. But old is not necessarily bad, at least for do-it-yourself (DIY) mechanics. These machines are easy to repair, and while brushes need periodic renewal, they should last for 600 or more operating hours. The AC sine wave these machines produce is fairly close to the grid wave, which is an important consideration.
- *Brushless generators* substitute electromagnetic induction for mechanical brushes. Brushless generators tend to weigh less than brushed generators for the power produced, making them a bit more portable. All major manufacturers produce brushless models, as do the makers of low-end imports.
- *Inverter generators* are the newest development made possible by the advent of high-amperage solid-state switches and powerful permanent magnets. The hardware is compact but sophisticated in function. Output appears as high-voltage, high-frequency AC, which is then rectified into direct current, filtered to remove ripple, and converted into 60-HzAC. Unlike brushed and brushless machines, which run at 3600 rpm, inverter generators turn only fast enough to satisfy load requirements. As a result, these generators are quieter, more fuel efficient, and have less of an environmental impact than conventional machines. But initial costs are high and the complex circuitry may be more than many DIY mechanics can cope with. Capacity is currently limited to 6 kW.

Table 1-1 lists specifications for some of the more popular generator models. Prices are very approximate and do not include shipping charges that may be involved.

Table 1-1
Specifications for some typical gensets

Make	Model	Type	Output power (W)	Output voltage	Engine	Weight (lb)	Noise level (dBA)*	Warranty	Price	Comments
Powerhouse (Northern Tool)	500Wi	Inverter	450 run, 500 max.	120 VAC/ 12 VDC	4-cycle OHV 37 cc	26.5	55/64	1 year consumer, 6 months commercial	$350	Limited application
Honda	EV20001	Inverter	1600 max., 2000 max.	120 VAC, 12 VDC	4-cycle 98.5 cc	49	53–59	3 years consumer and commercial	$1150	CARB§-certified, microprocessor-compatible
Yamaha	EF2000is	Inverter	1600 run, 2000 max.	120 VAC, 12 VDC	4-cycle 79 cc	44	59	3 years†	$1000	CARB-certified, microprocessor-compatible
Yamaha	EF2400isHC	Inverter	2000 run, 2400 max.	120 VAC, 12 VDC	4-cycle 171 cc	75	60	3 years†	$1500	CARB-certified, microprocessor-compatible
Wacker	GPi 4300	Inverter	3800W run, 4300W max.	120VAC	4-cycle, Robin, 265cc cid	99	na	5 years, 3 years on engine	na	High-quality Austrian engineering, manufactured in the United States

(continued on next page)

Table 1-1
Specifications for some typical gensets (*continued*)

Make	Model	Type	Output power (W)	Output voltage	Engine	Weight (lb)	Noise level (dBA)*	Warranty	Price	Comments
Gillette	GPN-50H	Brushless	4500 run, 5000 max.	120/240 VAC	Honda 270 cc	186	76	2 years commercial	$2000/ $2500	Ground-fault circuit interrupter (GFCI), volt and hour meters, alternator designed and manufactured in the United States, engine manufactured in the United States
MultiQuip	GAGHAB	Brushless	5000 run, 6000 max.	120/240 VAC	4-cycle 360 cc	184	79	2 years commercial	$2375	GFCI, voltmeter
Honda	EU6500 ISA-C (Calif. Model)	Inverter	5500 run, 6500 max.	120/240 VAC	4-cycle 389 cc	260	60	3 years residential and commercial	na	CARB-certified, oil shutdown, fuel gauge, remote start available
Powerhouse (Northern Tool)	9000	Brushless	7200 run, 9000 max.	120/240 VAC, 12 VDC	4-cycle	210	75	2 years consumer, 90 days commercial	$850	Less than 5 percent THD at 100 percent load

Table 1-1
Specifications for some typical gensets (*continued*)

Make	Model	Type	Output power (W)	Output voltage	Engine	Weight (lb)	Noise level (dBA)*	Warranty	Price	Comments
Generac XG Series 8000 W	005800-0 005747-0	Brushless	8000 run, 10,000 max.	120/240 VAC, 12 VDC	4-cycle, OHV, 410 cc	250	na	2 years residential, second year parts only, 1 year commercial	$1600	Designed and manufactured in the United States, fuel gauge, hour meter, low-oil shutdown

*Decibel (dBA) ratings should not be taken at face value.
†These small generators are rarely used in commercial service.
§California Air Resources Board

9

Generators produce alternating current that reverses itself 60 times a second in this hemisphere and 50 times a second in many other parts of the world. Engineers express frequency as hertz, abbreviated Hz. Thus, generators for the American market produce 60 Hz alternating current. Medium- and mid-sized generators provide both 120-VAC and 240-VAC to operate clothes dryers, air conditioners, and well pumps. And almost as an afterthought, several manufacturers offer 12-VDC power for battery charging. The direct-current output is far too dirty to power computers and other electronic devices.

If your generator fails, it will fail during a blackout when you need the power the most. Build quality is your assurance that this won't happen. In terms of quality, gensets fall into four broad groups:

- *Consumer.* Mid-quality machines, available from all large manufacturers in various wattages. Figure 1-5 illustrates one of the more popular units.
- *Contractor/commercial.* Workhorses used on construction sites and outdoor movie sets and favored by rental agencies (Fig. 1-6). Respected brands include, but are by no means limited to, Cummins-Onan, Dayton, Honda EB Series, MultiQuip, WINCO, and Yanmar. Most employ brushes.
- *Inverter.* Although costing twice as much per kilowatt as conventional gensets, inverter generators provide the clean, distortion-free power that digital equipment requires. Campers and home owners with modest power requirements make up the primary market.

FIG. 1-5 *A Generac residential generator, rated at 3250 W and adapted for propane use. The 20-lb tank provides 9 hours of half-load running time between refills.* Gempler's

FIG. 1-6 *The Gillette GPE 65H commercial-grade generator is a serious piece of American-made machinery. Gillette generator heads are also used on Dayton commercial-grade gensets and by the military.*

- *Bargain basement.* Imports with price as their only recommendation. Cynics suggest that you buy two for parts enough to keep one running. Manufacturing shortcuts include the use of copper alloy—rather than pure copper—windings and crimped, rather than welded, connections.

It also should be mentioned that some generators are standalone units, intended to be driven by belt rather than directly off the engine crankshaft. These two-bearing machines come out of an industrial tradition and are usually quite well made. DIYers with a spare engine lying about might look into this option.

Before buying something as complex as a generator, spend some time window shopping, just looking at the various machines, to get a sense of quality or its absence. You might also check out customer reviews that Amazon, Home Depot, and other large retailers publish on their websites. Note which generators are used on local construction sites, and make the rounds of tool-rental agencies. Rental service is a Darwinian test that only the fittest survive.

Read the warranty with care. A two-year limited warranty is the industry norm, although Wacker offers a five-year warranty on inverter generators. Gillette stands by its diodes and capacitors—the parts most likely to fail—into perpetuity. The cheapest imports provide 90 days of coverage.

Careful shoppers may want to check out http://wemakeitsafer.com/ Generators-Recalls for information on U.S. Consumer Product Safety Commission (CPSC) recalls that have totaled 300,000 units since 2005. Certain models from Poulan, Homelite, Husky, Black Max, North Star, and, surprisingly, Honda and Cummins are among those on the list. Most problems involve leaks from carburetors and plastic fuel tanks.

Power output

The single most important aspect of a generator is its power output, expressed in North America as watts or kilowatts (thousands of watts). A 6-kW machine promises to deliver 6000 W, or enough power to illuminate 60 100W light bulbs. In other parts of the world, output is expressed as kVA (kilovolt-amperes). For single-phase machines of the sort that nearly all home owners purchase, watts = volts × amperes, and these two ways of expressing power can be considered equivalent. A 12-kVA generator puts out 12 kW.

Advertised power ratings can be misleading, however. Generators carry three somewhat ambiguous power ratings:

- *Maximum power.* Good for perhaps 30 minutes and a 10 percent duty cycle. Some manufacturers confuse the issue by calling the maximum rating the surge rating.
- *Surge or starting power.* Tolerated for a few seconds.
- *Continuous power output.* Available 24 hours a day at a 100 percent duty. Even so, gensets live longer if loads are kept within 80 percent of the continuous rating.

For obvious reasons, generator makers like to describe their products in terms of the maximum rating, a measure that is well nigh useless for the average consumer. What we depend on is continuous power on tap as long as we need it. The surge rating comes into play when electric motors, uninterruptable power supplies, battery chargers, or welding machines first come online. You may notice the surge on grid power as the lights dim when a well pump or air conditioner starts. Consumer-level generators deliver 1.2 to 1.5 times more surge current than their continuous rating. The best construction-grade generators surge 2 to 2.5 times their normal output. Motor-starting loads may exceed generator surge capacities, but these short-term overloads are, up to a point, tolerated. Matching genset ratings to anticipated loads is described at the end of this chapter.

Power quality

Voltage

Nearly all gensets have some form of control to hold voltage steady under varying loads. Undervoltage dims lights and overheats motors. Overvoltage invites insulation failure and short circuits. Inverter generators provide the most precise voltage control, which is necessary to operate computers and other digital devices. Conventional brushed or brushless generators should have some form of automatic voltage regulation (AVR).

Frequency

Two-pole generators must turn 3600 rpm to produce 60-Hz current. A sensitive governor is required to hold this speed. Inverter generators bypass the rpm/frequency nexus and run only as fast as the load requires.

Distortion

Alternating current should have the form of mirror-image sine waves. Brushless gensets can distort these waves by as much as 40 percent. Some manufacturers seem proud of their engineering and provide distortion data. Others say nothing about it. If you run digital equipment, look for a genset with no more than 6 percent full-load distortion.

Engines or prime movers

According to the CPSC, light-duty gasoline engines used on generators have an operating life of 500 hours, which means that the average genset goes to the scrap yard within five to seven years of purchase. Field experience supports this rather dismal assessment.

Consequently, it is wise to purchase a generator with a serious engine, such as a Honda GX Series or a top-of-the-line Briggs or Kohler (Fig. 1-7). These engines all have overhead valves, cast-iron cylinder sleeves, ball-bearing mains, and forged crankshafts. And each has distinguishing features, such as Honda's aluminum push rods and Kohler's fuel injection. Briggs Vanguard engines are manufactured by Daihatsu, part of the Toyota complex, which is a good recommendation. Hatz, Yanmar, and Kohler Lombardi are the best choices for diesel power (Fig. 1-8). And speaking from experience, avoid Chinese-made Honda clones.

FIG. 1-7   *The Kohler Command Pro EFI is the only small industrial engine with closed-loop fuel injection and one of the few with hydraulic valve lifters. Other features include four-stage air filtration, full-pressure lubrication, rolling-element main bearings, and a forged alloy-steel crankshaft.*

Electric starting is a great convenience and pretty well mandatory for engines displacing more than 300 cc. Some manufacturers add to the convenience with remote-starting kits, which are also available on the aftermarket. Other important features are an accessible and reasonably priced oil filter, low-oil-pressure shutdown, a large steel—not plastic—fuel tank, and a reliable fuel gauge. The better engines have two or more stages of air filtration.

Fuel options

Unleaded gasoline is the default fuel for which most generator engines have been designed. However, as described in Chap. 2, gasoline poses safety and storage problems. Some owners opt for propane, also known as liquid petro-

FIG. 1-8 *Diesel engines, such as this Kohler example, promise better durability than gasoline engines and safer, more convenient fuel storage.*

leum gas (LPG). Although it is more expensive and less energy dense than gasoline, propane has an infinite storage life. It is also possible to operate a generator on natural gas, although adapting a movable generator to fixed plumbing can be a bit of a headache.

Diesel power is another option (see Fig. 1-8). Compression-ignition engines cost 30 to 50 percent more than their gasoline-fueled equivalents, but they should operate for decades without major problems. And diesel fuel, when treated with additives, tends to be quite stable.

Other features worth having

This section provides a list of desirable features. No genset has them all, and the relative importance of each depends on how you plan to use the machine.

Instrumentation

A wattmeter indicates how hard the generator works; frequency meters and voltmeters report how well the generator copes with various loads and are primary diagnostic tools. Contractor-grade gensets now have hour meters to alert the operator when maintenance is due. Of course, you can purchase these meters separately.

Circuit breakers

The main circuit breaker is an essential safety feature that cuts off all electric power. In addition, each receptacle should be protected with a circuit breaker that automatically trips in event of an overload. Well-designed machines often have additional insurance in the form of circuit breakers or fuses on internal circuits.

Ground-fault circuit interrupters

The U.S. Occupational Safety and Health Administration (OSHA) mandates that gensets used on job sites have ground-fault circuit interrupters (GFCIs) on 15- and 20-A, 120-VAC receptacles. A GFCI compares the amount of current going to the load with the amount returning via the return line. Unless a path to ground has opened, both current flows will be equal. A discrepancy of as little as 0.5 A causes the GFCI to open and disable the circuit. OSHA credits generator-mounted GFCIs with saving the lives of at least 1100 construction workers (Fig. 1-9). However, the protection is by no means absolute. A GFCI-protected generator can still hurt you.

Automatic idle control

When current output is low, the automatic-idle-control (AIC) feature permits brushed and brushless gensets to slow to two-thirds throttle, which saves fuel and reduces noise. Because the reduction in rpm results in a similar reduction in voltage and frequency, the AIC should be turned off when motors are started.

Noise

A generator should quietly go about its work so that people can almost forget it's there. Some gensets do just that with noise outputs of around 50 dBA, or roughly the same level as ordinary conversation. (dB is the abbreviation for *decibels*, and A means that the measurement is biased toward frequencies to which the human ear is most sensitive). Other models generate 70 dBA or more on the logarithmic decibel scale. This level of noise will have your neighbors knocking on the door at midnight.

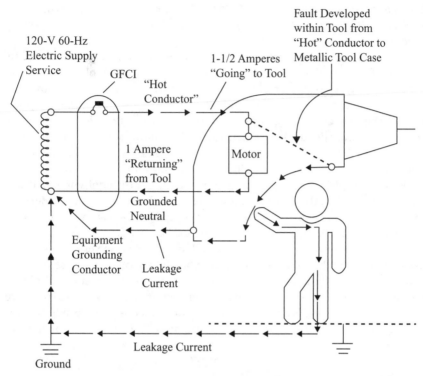

FIG. 1-9 *Without GFCI protection, a short to ground has lethal potential.* OSHA

Spec sheets can be misleading. Some manufacturers report decibel levels at full load with the meter positioned 3 meters (9.8 ft) from the machine. Others take the measurement at 7 m (23 ft) and at quarter load. Listen to a machine run before you buy it.

Sizing

A genset will supply only so much power, measured in watts, as the loads require. If the generator is unable to keep pace with demand, voltage drops, lights dim and flicker, and motors turn slowly, running on their starting windings. The engine labors and may stall. If the overload persist, motors and transformers burn out, and generator windings overheat, sometimes to the point of failure. This sort of damage is not covered by warranty.

There are two sorts of electrical loads—*resistive* and *reactive*. To determine the total resistive loads, merely add up the wattage of the electric heaters, light bulbs, stove elements, and so on—in short, whatever appliances use electricity to generate heat or light. (Fluorescent lamps are an exception.)

If you're running ten 100-W light bulbs and a 3000-W electric heater, the total load is 4000 W. You will need a genset rated at, say, 4500 W to comfortably meet this demand.

Switching on a motor, transformer, or fluorescent lamp is accompanied by a surge of current that the generator will try its best to supply. Motors draw especially large starting currents. How much depends on the horsepower of the motor, construction details, and how the load on the motor is apportioned. Fan and centrifugal-pump motors do not see their loads until they reach operating speed. Motors that drive air compressors and submersible well pumps go to work as soon as they are energized. Consequently, starting currents can be as high as eight times running current.

For most applications, it is enough to make a list of the equipment that will be powered and let the genset vendor determine the load. Factory dealers are experienced in these matters. For greater precision, you can have an electrician examine the equipment and total the loads.

Readers who want to make the load calculation themselves can use Table 1-2 as a very rough guide that whenever possible should be superseded by data from manufacturers' literature or equipment nameplates. If power requirements are expressed as amperes, multiply amperes by voltage to arrive at watts. However loads are calculated, err on the conservative side by purchasing 10 to 15 percent more generator capacity than you think you'll need.

Table 1-2
Typical loads

Equipment type	Running (W)	Starting (W)
Resistive loads		
75-W incandescent lamp	75	0
Fluorescent lamp		
1500-W space heater	1500	0
Water heater (40 gal)	4500–5500	0
Coffee maker, 4 cup	600	0
Electric range		
6-in. element	1500	0
8-in. element	2100	0
Reactive loads		
Microwave oven	700	1000
Refrigerator	700	2200
Freezer	700	2200
Clothes washer	1200	2400

Table 1-2
Typical loads (*continued*)

Equipment type	Running (W)	Starting (W)
Central air conditioner		
20,000 Btu	2500	3300
24,000 Btu	3800	5000
32,000 Btu	5000	6500
40,000 Btu	6000	7800
Window air conditioner		
10,000 Btu	1500	2200
Garage door opener		
1/4 hp	550	1100
1/2 hp	730	1400
Sump pump		
1/8 hp	800	1300
1/2 hp	1100	2150
Well pump, surface		
1/8 hp	800	1300
1/2 hp	1100	2150
Well pump, submersible		
3/4 hp	900	5400
1 hp	1200	7200
Drill		
1/8 in.	500	1000
1/2 in.	900	1800
Circular saw, 8 in.	1500	3000
Disk grinder, 7 in.	2000	4000
Air compressor		
3/4 hp	900	5400
1-1/2 hp	900	6000
Capacitor start	1800	10,800

2

Powering up

A portable generator is not a plug-and-play device. Producing one's own power requires an understanding of the safety issues, provision for fuel storage, and an appreciation of the tradeoffs involved in the ways the generator can be connected.

Location

Although long power cords are expensive (Table 2-1,) generators belong outdoors, as remote as possible from any dwelling, including those of the neighbors. Never run a generator in an attached garage, crawl space, or basement. If theft is a concern, chain the machine down, remove its wheels, or bring it in at night. If noise is a problem, purchase a quiet machine or modify the existing one with a better muffler.

According to Hal Stratton, chairman of the Consumer Product Safety Commission (CPSC), "The amount of CO from one generator is equivalent to [that of] hundreds of idling cars." And CO is deadly. The CPSC reports that between 1991 and 2011, at least 700 people died from inhaling generator exhaust. CO is an invisible, odorless gas that kills in minutes. Sleeping victims never awaken.

Some deaths are caused by gross ignorance, as was the case of the young English camper who placed a small handheld genset in his tent to keep warm at night. There are also records of people being treated in emergency rooms for acute CO poisoning from gensets that were 30 ft away from them. The risk depends on prevailing winds and the direction of the exhaust outlet. If you

Table 2-1
Maximum permissible ampere loads for
American Wire Gauge (AWG) power cords

Current draw (A)	50-ft AWG cord (A)	100-ft AWG cord (A)	150-ft AWG cord (A)
2	18	18	18
3	18	18	18
4	16	16	16
5	16	16	16
6	16	16	14
8	16	14	12
10	16	14	12
12	14	14	12
14	14	12	10
16	12	12	10
20	10	10	8

experience headaches, dizziness, nausea, weakness, or fatigue when in proximity—even outdoors—of a running generator, get to fresh air immediately.

Genset manufacturers warn about CO poisoning, but their manuals and sales literature often include images of gensets parked next to a house with a power cord passing through an open window or a cracked garage door. Nor have manufacturers been eager to use available technology to minimize the hazard. As of this writing, only two makers of marine gensets have installed catalytic converters that dramatically reduce CO levels in the exhaust.

The CPSC has demonstrated technology that automatically shuts down a generator when dangerous levels of CO accumulate (Fig. 2-1). But this technology has yet to be commeralized. One can, however, protect one's family with CO alarms in living spaces. Alarms should be battery powered and conform to Underwriters Laboratories (UL) 2034, International Approval Services (IAS) 6-96, or Canadian Standards Association (CSA) 6.19.01 standards.

Hookup

Connecting a generator can be as simple as plugging a lamp into a power cord or so complex that a building permit is required.

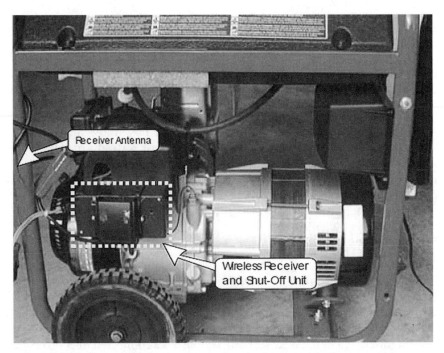

FIG. 2-1 *The CPSC staff fitted a universal radio frequency (RF) transmitter to a commercial CO detector and a receiver to the engine. When high CO levels trip the alarm, the transmitter signals the engine to shut down. Once shut down in this emergency mode, a delay circuit prevents the engine from being started for 20 minutes in order to provide time for the CO to dissipate.* CPSC

Power cords

Power cords passed through open windows or doors are the least expensive way of powering lamps and appliances that normally would plug into wall outlets. These cords are special items with only a distant relationship to ordinary extension cords.

Things to consider when purchasing power cords include

- *Safety.* Limit your choices to cords with National Electrical Manufacturers Association (NEMA) and/or UL listings. Cords for the Canadian market should be labeled as meeting CSA safety standards.
- *Connections.* There are dozens of female receptacles and matching plugs, all of them carrying a NEMA identification code. Purchase power cords with the correct plugs, as described in the owner's manual. Do not purchase plugs separately and make them up to power cords. Plugs must be factory installed and molded to the cord insulation.

- *Length.* The shorter the cord, the less resistance. However, mitigating the CO hazard requires that the generator be remote from any air intake into the house, including the cracked window or door through which the power cord passes. The location of appliances inside the house determines how much additional cord length will be necessary.
- *Gauge.* American Wire Gauge (AWG) numbers are counterintuitive in that the lower the number, the thicker the conductor and the longer the cord can be without imposing too great a power loss. A No. 12 cord carries more amperage than a No. 14 cord. A Husqvarna 50-ft 10/3 cord would be a good candidate for many applications.
- *Inspection.* Power cords see hard use, often in a wet environment. Abraded insulation or a missing ground pin can have tragic consequences.

Power inlet box

A power inlet box is merely an outdoor connection for the generator that obviates the need to run a power cord through an open window or cracked garage door (Fig. 2-2). An inlet box makes a neater job of the wiring, provides protection against weather, and reduces the likelihood of CO intrusion.

In nearly all installations, the output from the box goes to the transfer switch (discussed in the next section). People have connected inlet boxes to an array of indoor receptacles so that lamps and appliances can be plugged in without a tangle of extension cords on the floor. However, these receptacles must have no—zero—connection to existing household wiring. If you follow this example, you will end with two sets of receptacles: the existing set, connected to the grid, and the other set that has power only when the generator runs.

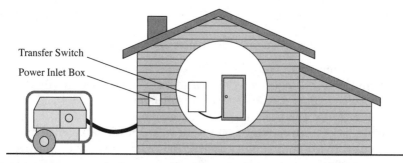

FIG. 2-2   *A power inlet box provides a central connection point for the generator.*

Transfer switches

A transfer switch shifts the load from the grid to the generator and, when power is restored, from the generator back to the grid. Installing one of these switches enables the generator to power household circuits without danger of backfeeding into the grid. Should backfeeding occur, the generator energizes the whole area, or "island," affected by the outage. Utility transformers boost the voltage to several thousand volts, which is both lethal to linemen and damaging to utility equipment. The danger of backfeeding is not academic—dozens of utility workers have been killed because someone powered up the lines with a home generator.

In order to positively prevent backfeeding, all transfer switches "break before make." That is, the switch breaks continuity with one power source before it makes contact with the second source:

LINE—Utility power ON, generator power OFF.
OFF—Utility power OFF, generator power OFF.
GEN—Utility power OFF, generator power ON.

From the point of view of the home owner, the beauty of a transfer switch is that it enables existing circuits to be powered as if the grid were functional. One no longer needs to clutter the floor with extension cords. Hardwired loads, such as central air conditioners, furnace fans, and well pumps, also can be powered by the generator.

Figure 2-3 illustrates a manual transfer switch, which must be thrown by the home owner. Automatic transfer switches (ATSs) require no human intervention. When grid power drops to a predetermined level, the ATS automatically connects the generator to household wiring and disconnects it when grid power is restored. Gensets that are permanently parked and poised to run can be started and stopped automatically by signals from the ATS. These switches can be programmed to exercise the generator at predetermined intervals, announce when maintenance is due, and give priority to important loads. When a central air conditioner or a well pump starts, the transfer switch senses the reactive load and temporarily shuts off power to other circuits.

The *National Electrical Code (NEC)* and local ordinances require that a certified electrician install a transfer switch whenever generator power could enter distribution lines. In addition to having a certified electrician perform the work, most jurisdictions require a permit and follow-up inspection for transfer-switch installations. The local utility also must be notified. Moreover, transfer switches are not cheap; one can spend $1000 or more for a state-of-the-art ATS.

Some home owners avoid the cost of a transfer switch and three or four hours of electrician's labor by installing a mechanical lockout on the main

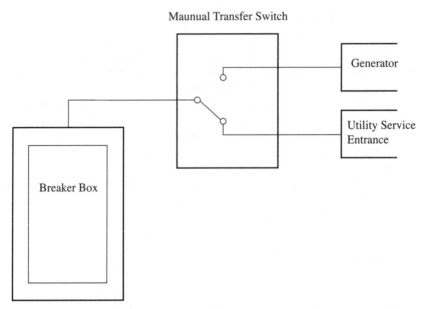

FIG. 2-3 *A manual double-pole, double-throw (DPDT) transfer switch. Double pole means that it switches two hot leads. The neutral lead passes uninterrupted through the switch. Double throw refers to the two power-related switch positions—OFF is not counted.*

power panel. The *NEC* and many local codes do not accept these lockouts. If something goes wrong, the lockout will be plaintiff's exhibit No. 1.

Transfer-switch installation options A transfer switch can be installed either between the meter and the main distribution panel (breaker box) or downstream of the panel (Fig. 2-4). With a large enough generator, the former option can power the entire house (the typical American home draws 60 A or more). Some owners prefer this *service-disconnect option* even when the intent is to power only those circuits that are considered critical. The initial cost of such a setup is higher because the amperage of the transfer switch must be rated the same as the service amperage. On the other hand, a sub-panel is not needed.

Most people find the *partial-service option*, where the transfer switch mounts downstream of the main distribution panel, more practical. The transfer switch needs only to be rated at generator amperage, and connections to the main panel can be made without having the utility cut off line power. One merely opens the main breaker. On the other hand, a subpanel is required to carry breakers for the circuits that the generator powers. If the subpanel

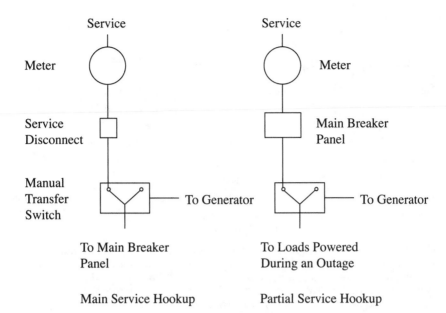

FIG. 2-4 *Two methods of connecting a transfer switch. The main service panel is also known as the main distribution panel, the load center, or more informally, the breaker box.*

shares the same enclosure as the transfer switch, the unit is often described as a *transfer panel* or *generator panel*.

Transfer switches and neutral grounds

Dual-voltage (120/240V) gensets may have the white or gray neutral conductor grounded, or bonded, to the genset frame, or the conductor may be unbonded, or floating. The genset should be labeled accordingly.

This technical detail is important because it determines how the genset should be connected to household wiring. The presence of a ground-fault circuit interrupter (GFCI) on one or more 120 VAC genset output circuits further complicates matters. A bonded genset with GFCI protection will not function when connected to a conventional two-pole transfer switch. As soon as the switch is turned to GEN, the GFCI trips and shuts off output current.

Two options are available to put the generator into service:

- An electrician can convert the generator to a floating, or unbonded, neutral and use a two-pole transfer switch (Fig. 2-5). When this modification is made, the *NEC* requires that the label "Neutral Bonded to Frame" on the genset be replaced with a label stating "Neutral

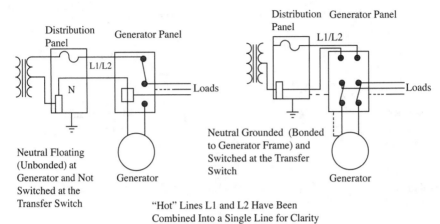

Distribution Panel — Generator Panel — L1/L2 — N — Loads — Neutral Floating (Unbonded) at Generator and Not Switched at the Transfer Switch — Generator

Distribution Panel — Generator Panel — L1/L2 — Loads — Neutral Grounded (Bonded to Generator Frame) and Switched at the Transfer Switch — Generator

"Hot" Lines L1 and L2 Have Been Combined Into a Single Line for Clarity

FIG. 2-5 *A two-pole transfer switch merely switches L1 and L2 "hot" conductors. The neutral conductor passes solidly through the switch on its way to the generator. A three-pole transfer switch makes and breaks all three conductors. A two-pole switch can be used with an unbonded, or floating, neutral, as shown on the left. Gensets with a bonded neutral require a three-pole switch.*

Unbonded." Because this modification disables the GFCI, the genset will no longer pass an Occupational Health and Safety Administration (OSHA) job-site inspection.

- The electrician can leave the neutral bond in place and specify a three-pole, neutral-switching transfer switch. This option is code legal but frowned on by electrical inspectors. No portable generator lasts forever, and the replacement may have an unbonded neutral, which would be illegal to use with the three-pole switch [*NEC* 250, 20(B)(1)].

WARNING: Consult with the dealer or manufacturer before attempting or allowing any modifications to a genset, particularly inverter models.

Electrical safety

The CPSC has worked hard to convince the public that portable generators should be placed outdoors, remote from dwellings. At the same time, the CPSC recognizes that outdoor locations pose risk of electroshock. As a CPSC memo put it:

Presently, portable gasoline-fueled generators are not constructed to permit their use outdoors during inclement weather. This creates confusion for users, who cannot place their units indoors because of significant levels of carbon monoxide emitted by the gasoline engine. Weatherproofing portable-engine generators appears to be an attractive strategy to eliminate the confusion and potential for electroshock associated with outdoor use.

This memo was written a decade ago, and nothing has changed. It's up to owners to keep their generators dry, which is admittedly a difficult proposition in a hurricane. A canopy-like cover with ample ventilation for engine cooling is one option. Elevate the generator high enough to avoid flooding, and try to make yourself as least conductive as possible. Do not stand in water while servicing the generator, wear rubber-soled boots, and dry your hands before touching the power cord or generator frame.

Grounding rods

Owner's manuals insist that all gensets should be grounded with copper rods driven several feet into the ground. OSHA agrees that *grounding rods should be used with most gensets that deliver power to homes, recreational vehicles (RVs), trailers, and other structures* through a transfer switch. But OSHA permits the generator frame to act as ground if the system has no permanent connection with the grid.

To understand this OSHA's position, we have to detour a moment to the *NEC*. In the *NEC*'s scheme of things, the way the transfer switch handles the ground conductor determines whether the system is separately derived (completely isolated from the grid) or non-separately derived (sharing a common ground with the grid):

- *Separately derived system.* If there is no transfer switch, that is, if tools or other loads plug in directly to the generator or connect to it via extensions cords, the system has no grid connection and is separately derived. Generators that power outdoor lighting or that are used on construction sites are examples.

    The system is also separately derived if the generator connects to a transfer switch that makes and breaks the neutral ground connection. Typically, single-phase, 3-wire and three-phase, 4-wire transfer switches fall into this category.

- *Non–separately derived system.* Most homeowner applications are non-separately derived, since associated transfer switches do not open or close the ground connection. The system ground remains "solidly connected" to the grid ground.

The omission of grounding rods applies only to the separately derived systems that have generators with the following characteristics:

- The generator powers only equipment mounted on the generator and/or through extension cords that plug in directly to receptacles on the generator frame [*NEC* 1926.404(f)(3)(i)(A)].

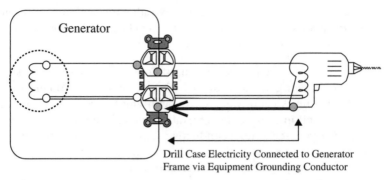

Drill Case Electricity Connected to Generator
Frame via Equipment Grounding Conductor

FIG. 2-6 *Tool frames must ground to the generator frame.* John "Grizzy" Grzywacz, Professor Emeritus, OSHA National Training Institute

- The fuel tank, engine, generator housing, and other metal parts of the genset are bonded to the generator frame. *Bonded* means that connections are electrically conductive [*NEC* 1926.404(f)(3)(i)(B)].

In addition, generator receptacle grounding pins and the grounding wires from tools and other loads must be grounded to the generator frame (Fig. 2-6).

A grounding rod actually increases the likelihood of electrocution (Fig. 2-7). The rod creates an alternative current path from an electrical fault to the tool case, through the operator to ground, and back to the generator. About all that can be said for grounding rods is that they give protection against lightning strikes.

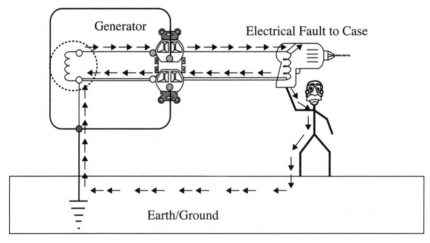

FIG. 2-7 *A grounding rod can have lethal consequences if a tool or other piece of equipment develops a short circuit.* John "Grizzy" Grzywacz, Professor Emeritus, OSHA National Training Institute

Fuels

A generator is not much use without fuel to operate it. Unless you rely on natural gas, some provision for fuel storage must be made. Gasoline stations will be closed during a blackout. Fortunately, most outages are short-term affairs caused by accidents or routine maintenance. According to the 2011 Eaton Blackout Tracker, short-term outages, on average, last about 3-1/2 hours. California experienced the most outages, with 371 during the year, followed by New York and Texas.

Outages induced by severe weather are another matter. Every hurricane, tornado, ice storm, or *derecho* is unique. Generating and distribution systems vary in vulnerability and their responsiveness. Utilities share repair crews and hire contract workers during emergencies. But days can pass before these extra hands are deployed and no work can be done until wind speeds drop to 30 mph. Hurricane Hugo knocked out power for 18 days while 8608 transformers and 700 miles of transmission lines were replaced. Hurricane Fran left more than 1 million people without power for as long as 10 days. From the fragmentary data available, it appears that the typical large-storm blackout lasts for about six days.

Most genset manufacturers describe fuel consumption in terms of hours between refills at half-throttle. Honda and a few other Japanese makers provide upfront gallons-per-hour data. For example, Honda's EB6500 gasoline genset consumes 0.9 gal/h at its rated 6500-W load. This translates to 130 gal consumed during a six-day period. Even if the generator runs for 6 of every 24 hours, you would still need to have 30 gal of fuel on hand.

Gasoline

Nearly all gensets operate on unleaded gasoline, the most readily available fuel and the one for which most engines were designed to use. As owner's manuals warn, gasoline supplies should be stored in a shed at some distance from the house and never in an attached garage or basement. Shut the engine down and wait 10 minutes for it to cool before refueling. Wipe up any spills, and allow time for the residue to evaporate before restarting the engine. Do not top off the tank because an inch or so of free space must be present to allow for fuel expansion. Move the fuel can well clear of the engine before starting, and have an ABC fire extinguisher within arm's reach.

Long-term storage is a problem because gasoline oxidizes and should be used within 60 days of purchase. Storage time can be extended by adding a fuel stabilizer, such as Briggs & Stratton's Advanced Formula Fuel Treatment & Stabilizer. The company claims that the product can extend fuel life for as long as three years. If necessary, you can transfer gasoline that's approaching its due date to your car.

The safest gasoline containers carry an OSHA stamp of approval. This means that the container holds no more than 5 gallons and has a spring-loaded lid, a flame-arresting screen, and a provision for venting in event of fire. In addition, the container must be approved by the U.S. Department of Transportation (DOT) and certified—not merely listed—by UL or another established rating agency. Most OSHA-approved containers are made of steel, although the agency does approve of polyethylene containers that meet its requirements. Note that OSHA-approved standard plastic containers have no resemblance to the more ordinary types, which leak vapor and spill fuel out of their extensible spouts.

Propane

Gasoline engines can be factory or field modified to burn propane, a gaseous fuel that has an indefinite storage life. Propane burns cleaner than gasoline and requires less frequent oil changes. Barbeque-pit 20-lb tanks are the norm, but propane can be purchased much more cheaply if the owner has a large storage tank, which can be leased or purchased outright (Fig. 2-8).

Propane cylinders, or "bottles," develop 200 lb/in.^{2} of pressure on a warm summer day. Most are made of welded steel and, for consumer use, mount

FIG. 2-8   *20- and 100-lb propane containers.* City of New York Fire Department

upright to ensure that vapor, not liquid propane, is discharged. Cylinders have a service valve, a relief valve that automatically opens to vent excessive pressure, and permanent identification, including the serial number and date of manufacture. Cylinders must be visually inspected by certified personnel 12 years after the build date and every 5 years subsequently. Tank size is nominated in pounds, although the relationship to weight is tenuous at best (Table 2-2).

A regulator reduces line pressure because engines cannot tolerate a 200-psi fuel stream. Regulator hardware has left-hand threads, tightened by turning counterclockwise. Figure 2-9 illustrates potential leak points. Test with soapy water—never by use of an open flame. A spray dispenser set to deliver a solid stream is the tool of choice. Leaking propane makes itself known by

Table 2-2
Fuel capacities of small propane tanks

Tank size	Capacity (gal)	Weight, full (lb)
No. 20	4.7	39
No. 30	7.1	54
No. 40	9.4	70
No. 100	23.5	170

FIG. 2-9 *Using soapy water, check the X-ed joints for leaks.* City of New York Fire Department

the rotten-egg smell of mercaptans introduced during processing. Mount the bottle upright at least 10 ft from the generator, and connect with a DOT-approved 300-psi hose.

Natural gas

Natural gas is an attractive option, especially since generators set up for natural gas can be adapted to run on gasoline or propane. Unless there is an earthquake or the pumping station floods, natural gas will be available when you need it. On the other hand, natural gas varies in quality. Check with your local utility to verify that the water content and delivery pressure are compatible with engine operation. Depending on your skills, professional help may be needed to anchor the genset properly and make up a safe hose connection to the gas line.

Diesel

Diesel is the most energy-rich fuel and probably the safest to use and store. Number 2 diesel fuel has a shelf life of about 18 months and can be conveniently stored in 55-gal drums. Various products are available to combat oxidation and algae formation.

Note that unlike gasoline engines, diesel engines should not be run dry in preparation for storage. Drain the tank, but leave fuel in the lines. Otherwise, you will have to purge air from the fuel system before starting. Nor do injection pumps take kindly to water. Never refuel diesel engines in the rain.

Planning ahead

Sooner or later, grid power will go down. And it may stay down for hours, days, or even weeks. The best assurance that the generator will start is to run it briefly at half load once a month. It's also a good idea to have an emergency kit stocked with consumable parts, such as spark plugs and air, fuel, and oil filters. Some gensets require an oil and filter change every 25 operating hours. Chapter 3 includes a list of tools and supplies that will be needed if something goes wrong with the machine. Store emergency tools and backup parts together so they can be found in the dark.

Most difficulties that arise from long-term power outages are human. To cope, a family will require

- Bottled water (1 gal per person per day)
- Nonperishable food for humans and pets
- Can opener
- Prescription drugs

- First-aid kit
- Matches
- Flashlights, one with a hand crank
- Oil lamps with an adequate supply of lamp oil
- Fresh batteries for flashlights, smoke and CO alarms, radio, and clock
- Fire extinguisher
- Cell phone (inverter to recharge the cell phone from a car battery in event the generator fails)
- Weather radio (battery operated)
- Extra clothes
- Blankets
- Games for children, books, and other ways to pass the time

3

Generator repairs

Some 300 brands of portable generators, mostly of Chinese origin, are on the market. Even if the data were available, it would be impossible to provide step-by-step repair procedures for each of these machines.

Fortunately, generators fall into three broad families—brushed, brushless, and inverter. Once you understand what brushed generators must have to function, you can usually fix any of them. The same holds for brushless types. Inverter generators present some difficulties that will be addressed later.

Another factor working in favor of do-it-yourself (DIY) mechanics is that most genset suppliers assemble rather than fabricate, their products. That is, generator heads, engines, and control circuits are off-the-shelf components, which imposes a degree of standardization on the industry and simplifies the search for replacement parts.

Honda, Briggs & Stratton, Gillette, Generac, and a few other mainline manufacturers include wiring schematics in the owner's manuals or make them available on the Internet. Others do not either because of slap-dash customer support or as a means of keeping service information proprietary. Clymer's two-volume *Small AC Generator Service Manuals* are a good but necessarily incomplete and dated source of wiring schematics and other data. I have included as many representative diagrams here as space permits, but unless you have access to the specific diagram for your machine, you'll be navigating without a map.

Working "blind" is not as daunting as it might appear. Most failures involve a handful of components that can be isolated and tested. For example, failure to produce voltage usually involves the diodes, the capacitor (a cylinder about the size of a large salt shaker), the automatic voltage control box, or the brushes (when present). Color codes assist in tracing the wiring from the outlet receptacles, through the GFCI and back to the stator windings.

The more you know about theory, about how electricity behaves and the function of the various genset components, the easier troubleshooting becomes. Those who work without a theoretical backdrop end by throwing parts at the problem, which is an expensive and time-consuming exercise.

Most of the procedures described here are quite safe because they are carried out with the generator stopped. You might get zapped by a capacitor that you have neglected to discharge, but that's about the extent of personal risk.

Other test procedures involve measuring voltage on internal parts while the generator runs. If you are new to electricity or uncomfortable working around lethal voltages and rotating machinery, skip these tests or have someone more proficient make them. A genset is not a teaching aid.

Basic safely precautions include:

- Disconnect the generator from household wiring.
- Check for shorts to ground before touching metal parts on any defective or recently repaired generator.
- Disconnect the battery on electric start machines.
- When working on a running generator, always have someone standing by to shut off the machine.
- Wear rubber-soled shoes and, when practical, insulating gloves.
- Inspect test lead insulation, probes, and alligator clips to verify that they will withstand the levels of voltage encountered.
- Be especially careful with wet-cell, automotive batteries. These batteries give off hydrogen gas that can ignite in the presence of sparks. Use long charging leads and cover the battery.
- Work in a well-ventilated area so that exhaust fumes can dissipate.
- Cap and store fuel containers at a safe distance from the work site.
- Wipe up fuel spills and repair any fuel leaks before starting the generator.

Conventional generators

At their most basic level generators depend upon Faraday's law of induction, which states that exposing a conductor to a moving magnetic field induces a voltage in the conductor (Fig. 3-1).

Nearly all portable generators are synchronous machines that take output from the stator. These generators consist of a rotor, a stator with copper windings that react to rotor magnetism to produce output current, a cooling fan, and a bearing that supports the outboard end of the rotor and keeps it cen-

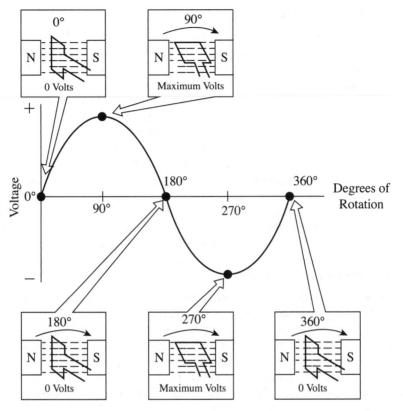

FIG. 3-1 *An alternating-current (AC) generator or alternator produces voltage as a conductor moves through a magnetic field. The intensity of the voltage peaks with each half rotation and reverses direction during the next half rotation. The rise and fall of voltage (and current) take the form of mirror-image sine waves with an intensity determined by the strength of the magnetic field and conductor velocity. Frequency is a function of rotor rpm. This drawing shows voltage as coming off the rotor and the magnet in the stator, or stationary part of the machine. Inverter generators work in this manner. Conventional brushed and brushless generators carry the field magnets in the rotor and take current off the stator.* DOE FSC-6910

tered within the surrounding stator. The inboard end of the rotor is located by the engine main bearings.

The rotor, or armature, consists of heavy-gauge copper coils wound over a laminated sheet-iron core. Two-pole rotors, which are the industry standard, have two magnetic poles and must be turned 3600 rpm to induce 60-Hz current in the stator. Four-pole, 1800-rpm rotors are found in heavier, often diesel-powered gensets.

Brushed generators

Brushed generators are workhorses, favored for hard service. Every manufacturer of note, including Briggs & Stratton, Coleman, Dayton, Gillette, Honda, Wacker, and Yamaha, produce them. Outputs are clean enough to be used with portable tools and most household appliances.

You can recognize these generators by the presence of two carbon brushes that bear against slip rings on the outboard end of the rotor (Figs. 3-2 and 3-3). Each slip ring is insulated from the rotor shaft and connected to one of the two main rotor windings. The heavy-gauge copper windings, imbedded in slots on the rotor, are 180 degrees apart with clockwise current flow for the south magnetic pole and counterclockwise flow for the north pole. When excited and put into motion, rotor windings induce current in the two main stator windings, which is the purpose of the exercise. Power for excitation comes from the stator, undergoes full-wave rectification, and passes into and out of the rotor windings through the positive and negative brushes.

Brushless generators

Brushed and brushless generators differ in the way excitation current is transmitted to the rotor (Fig. 3-4). Brushed machines use a mechanical connection; brushless generators depend on magnetic induction created by the interaction of the stator exciter winding with the rotor windings. Diodes mounted on the rotor rectify the induced AC, converting it to the direct current required to excite the main rotor windings. Brushless generators also have a large capacitor in series with the excitation circuit to provide voltage control.

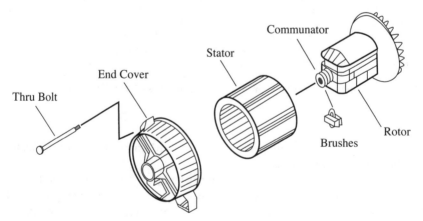

FIG. 3-2 *Briggs & Stratton brushed generators are simple, easy-to-repair machines for which spare parts are readily available.*

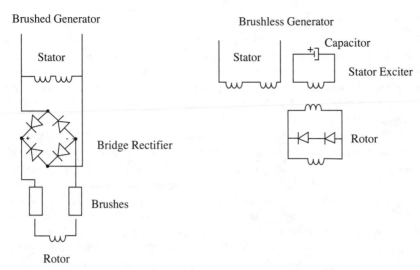

FIG. 3-3   *Excitation current for brushed generators passes through the brushes and slip rings. Brushless generators induce alternating current into the rotor through an exciter circuit. Rotor diodes then rectify the AC, converting it to direct current.*

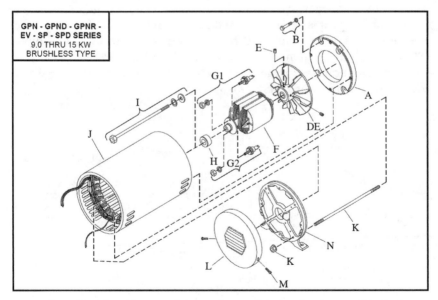

FIG. 3-4   *Gillette brushless generator end cover (A), fan (D,E), rotor (F), diodes (G1, G2), ball bearing (H), stator (J), and bearing casting (N).*

Power quality

Voltage and frequency vary with load primarily because loads form such a large fraction of genset capacity and secondarily because gensets lack the rotating mass to hold rpm steady. A sudden load bogs the engine with corresponding voltage and frequency dips (Fig. 3-5). The problem cannot be attacked directly; heavy flywheels and oversized prime movers would put an end to portability.

Until recently, genset owners with a need for voltage and frequency stablity had to take it on themselves to modify their machines with industrial-quality speed sensors, fuel pumps (to replace gravity feed), and steeper motors to give precise control over throttle angles. Manufacturers have since improved engine governors enough to make these modifications redundant. But the real breakthrough came with digital automatic voltage regulation (AVR) applied to both brushed and brushless generators. When a load comes on line, a silicon-controlled rectifier (SCR) in the AVR responds to the voltage dip by increasing current flow in the excitation circuit, which intensifies rotor magnetism and stator output. Under reduced load, the SCR acts in the reverse manner, decreasing current flow to the rotor.

An AVR senses voltage, which works fine for resistive loads. But the inductive loads associated with motor starting, uninterruptable power supplies (UPSs), and other digital hardware require a transformer, either standing alone or in tandem with the AVR (Fig. 3-6). Well-designed generators hold voltage flat and hold frequency within 1 percent or less of 60 Hz (Fig. 3-7).

Generator output is also subject to distortion in the form of harmonic frequencies, which are multiples of the fundamental frequency. For example, 180 Hz is the third harmonic of 60 Hz, and 240 Hz is the fourth harmonic. The resulting waveform has a jagged, saw-toothed shape that can damage digital equipment and overheat generator windings.

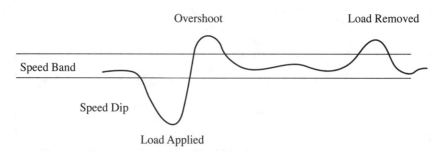

FIG. 3-5 *Dither—the loss of rpm under load followed by an overcorrection—is characteristic of mechanical governors.*

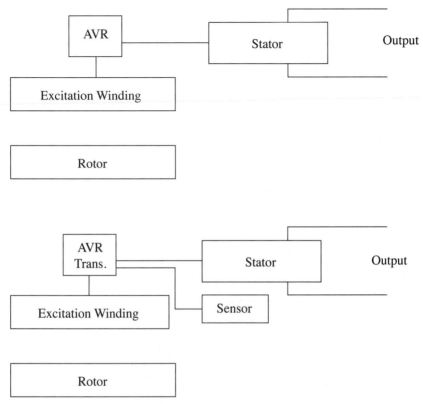

FIG. 3-6 *AVRs control generator output by regulating the rotor exciter circuit. In the upper drawing, the AVR senses voltage from the main stator windings as an indicator of load on the generator. In the lower drawing, a dedicated winding inputs load data into the AVR, which in this example is supplemented by a transformer. The exciter circuit also can include other components, such as a choke coil and a current-limiting varistor.*

Much of the difficulty originates with the generator, especially brushless generators with capacitor voltage control. Some of these machines, and not necessarily the least expensive models, exhibit 40 percent total harmonic distortion (THD). Gillette brushless generators do considerably better, with TDHs of 6 percent.

The type of load also affects the cleanliness of the sine wave. Computers, uninteruptable power supplies, precision motor controls, and digital sound systems are among the worst offenders, as are fluorescent lamp ballasts. Many of these devices malfunction if THD exceeds 10 or 15 percent. As a point of reference, utilities try to keep THD within 3 percent.

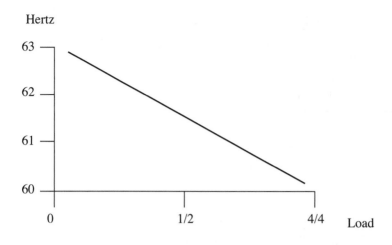

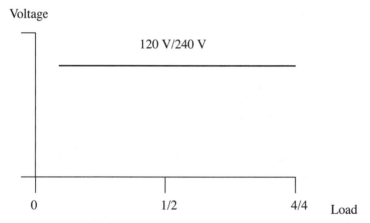

FIG. 3-7 *Frequency and voltage curves for the Robin American RGV12000 demonstrate the power quality a well-designed brushed machine can deliver.*

Inverter generators (discussed later in this chapter) eliminate generator-induced harmonics and, at their best, provide almost perfect voltage and frequency control.

Tools

Electrical repairs require

- *Digital volt-ohmmeter (VOM)*. A volt-ohmmeter, or multimeter, is the first tool a technician reaches for. The instrument should have these capabilities:

- AC voltage: 0–250, 0–500 V
- DC voltage: 0–20 V
- Amperes: 0–10 A
- Resistance: 200–2kΩ, 20kΩ–200kΩ, 2mΩ–20mΩ

The Triplett Model 9007 retails for around $60 and includes a capacitor tester. The $140 Fluke 114 is another good general-purpose multimeter that measures the usual variables plus capacitance, frequency, and diode function.

- *Clamp-on ammeter.* This instrument senses current flow inductively through a pair of calipers that snap over the conductor. The Fluke i200 clamp-on ammeter accessory extends the range of the 114 Multimeter to 200 A. You also can use a conventional ammeter that must be connected in series with the circuit. In either case, desirable amperage ranges are 0–10, 0–50, and 0–100 A.

- *Tachometer.* The Matco TA-100 and the Echo PET-304 sense engine speed from voltage spikes in the spark plug lead. The $35 Harbor Freight Cen-Set measures rpm with reflective tape and an infrared light source. The vendor claims ±1 rpm accuracy at speeds above 1000 rpm. Of course, a rotating element, such as the flywheel, must be visible for the tool to function.

- *AC frequency meter.* Many DIYers judge frequency by engine rpm, but a professional-grade hertz meter gives far more precise results. One also might consider a Kill-A-Watt, an inexpensive little tool that also reports voltage and amperage (Fig. 3-8).

FIG. 3-8 *A Kill-A-Watt reports voltage, frequency, and amperage, although the amperage function is limited.* Courtesy of Kill-A-Watt

- *Load center.* If generator repairs are to be more than hit or miss, some means of progressively loading the generator must be available (Fig. 3-9). Electric room and water heaters can provide the bulk of the load, with incandescent lamps taking up the slack.
- *Soldering gun.* A soldering gun, 18-gauge rosin-core solder, and heat-shrink tubing are essential. Cheap generators use coarse, multistrand wiring that work hardens with vibration and breaks. Although most suppliers don't stock it, fine multistrand replacement wire of the sort used for VOM probes is worth looking for.

A megaohmmeter (megger) and an oscilloscope are useful but prohibitively expensive tools. Most DIY mechanics make do without them, although vintage, hand-cranked meggers sometimes can be purchased on Ebay for a few dollars. Used oscilloscopes are rarely worth the money.

Measurements

Figure 3-10 illustrates how amperage, voltage, and resistance are measured.

Output tests

Verify that the generator turns 3750 rpm (±50 rpm) under no load, a speed that should produce anywhere from 117 V (for light-duty units) to 125/130 V

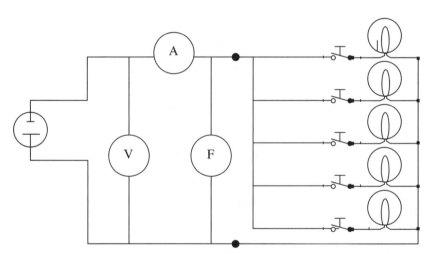

FIG. 3-9 *A load center consists of purely resistive loads that, in this example, are in the form of incandescent lamps.*

Genset Running

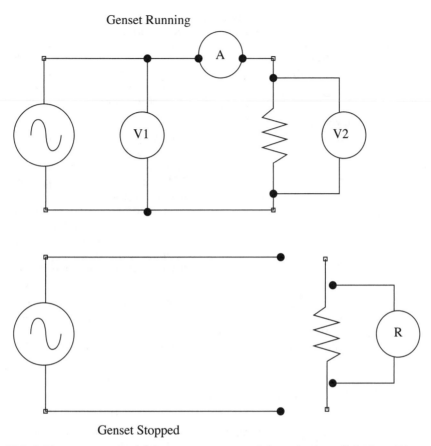

Genset Stopped

FIG. 3-10 *An ammeter (A) connects in series, a voltmeter in parallel. Meter V1 reads the voltage across the entire circuit; V2 reads the voltage drop across the load. Resistance measurements (R) are made with the power shut off and the component under test disconnected.*

at 61 or 64 Hz. You might have to adjust the governor to achieve the correct no-load engine speed.

Observe the effects of load on rpm and output as you add loads (Table 3-1). At full rated load, the generator should be good for 118 V or so at between 58 and 60 Hz. A marked loss of rpm suggests an engine problem. If engine speed drops 10 or 15 percent (the figure varies with the manufacturer), the generator will shut down to protect the loads from low voltage. A less-than-perfect correlation exists between voltage irregularities and generator faults.

Table 3-1
Output voltage as a diagnostic tool

Without load	0 voltage on a single receptacle, others with normal voltage	1. Faulty receptacle or ground-fault interrupter circuit (GFCI) 2. Bad connection 3. Break in wiring to receptacle
	0 voltage on all receptacles	1. Faulty main breaker 2. Loss of residual magnetism 3. Worn or sticking brushes 4. Shorted stator windings 5. Blown voltage-control capacitor 6. Faulty AVR or improperly adjusted voltage trim pot on AVR 7. Low engine rpm that trips the underfrequency protection circuit to shut down output 8. Loose mechanical connection at the generator roter and engine crankshaft
	Half-normal voltage (60 V at 120 V receptacles, 120 V at 240 V receptacles)	1. Faulty bridge rectifier or rectifier connections 2. Faulty or misadjusted AVR 3. High-resistance connection in output circuit
	Less than half to three-quarters voltage	1. Blown voltage-control capacitor 2. Faulty or misadjusted AVR
	8 V to 12 V on all receptacles as the result of residual magnetism	1. Fault in rotor exciter circuitry 2. Shorted stator windings
	Excessive voltage on all receptacles	1. Engine speed too high 2. Faulty or misadjusted AVR 3. Faulty AVR sensor-circuit wiring 4. Bad sensor-circuit transformer (when present) 5. Bad sensor-circuit current limiter (when present) 6. Overly large voltage-control capacitor installed
With load	Fluctuating voltage	1. Faulty AVR 2. Misadjusted AVR stability pot 3. Low stator/rotor winding resistance to ground 4. Worn rotor-shaft bearing

Table 3-1
Output voltage as a diagnostic tool (*continued*)

With load (*continued*)	Voltage falls sharply as loads are imposed.	1. Open stator 2. Defective diodes (that may pass resistance tests) 3. Faulty AVR 4. Misadjusted AVR stability control pot 5. Unresponsive engine governor 6. Poorly performing engine 7. High current draws imposed by loads with a power factor of less than 0.7 8. Excessive loads
	Persistently low voltage under load	1. Engine speed droop excessive—check engine and governor 2. Faulty AVR 3. Faulty AVR power or sensing circuit 4. Faulty rotor diodes 5. Faulty rotor winding 6. Excessive resistance in power cord
Noise	Worn rotor-shaft bearing	1. Check for contact damage on the rotor and stator

Flashing Conventional (brushed and brushless) gensets bootstrap themselves during starting by virtue of residual magnetism that lingers in the stator exciter winding or by means of a small permanent magnet imbedded in the rotor. With either configuration, the generator must develop 5 to 8 V noload to start.

Gensets that depend on residual magnetism can, if idled for long periods, require flashing to restore the magnetic field. There are various ways to do this, none of which are absolutely safe for the equipment or the operator. Readers new to the subject should farm out the job—which takes no more than five minutes—to a professional. Once you see how it's done for your machine, flashing the generator yourself presents no problem.

Overspeeding Manually override the governor to run the generator under full load at 4000+ rpm for two seconds. Sometimes the effect of the higher speed on what little residual magnetism remains brings the generator back to life.

Drill motor   This procedure, said to be recommended by Briggs & Stratton, is another hit-or-miss approach. But it's worth a try:

- Plug a 1/4-in. electric drill into a generator 120 V receptacle.
- Start the generator.
- Spin the drill chuck counterclockwise by hand while holding down the trigger. If this doesn't work, spin the chuck in the other direction. Rotating the chuck converts the motor into a DC generator.

CAUTION: The drill will start once 120 V comes online.

Battery-to-rotor   There is no universal one-size-fits-all way to impose battery voltage on the rotor. What one manufacturer recommends damages another manufacturer's product (Table 3-2). What these procedures have in common is that the voltage transfusion enters the rotor by way of the brushes on brushed models and by way of the capacitor on brushless machines. Also note that many of these procedures are accomplished with the generator running, its end cover removed, and your fingers in proximity to hot wires and moving parts. Wear insulating gloves.

Resistance tests

WARNING: Generators that start automatically in the absence of grid power or that can be started with a remote switch must be locked out before performing this and other service operations.

From the point of view of the equipment and of the person working on it, the safest way to diagnose generator faults is to make resistance tests on individual components. The generator must not be running, and each component must be disconnected from its associated circuitry. This somewhat laborious procedure—handling each part, tracing out the circuits—makes the technology understandable in ways that verbal descriptions cannot.

Where one starts is a matter of choice. The approach used here is to test the GFCI, the main breaker, and output receptacles first. These components fail fairly often. Stator and rotor tests follow because the costs of replacement parts can make further repairs moot. Brushes, AVRs, rectifiers, and other components that make up the exciter circuit are the last to be tested.

GFCIs   Ground fault circuit interrupters are complex devices that cause more than their share of problems. If the GFCI continually trips, assume that it's doing its job and cutting off power in response to a short in the generator or load. It's also possible that vibration, intensified by mismatched loads, worn

Table 3-2
Flash procedures for some popular gensets

Make	Model	Flashing procedure with generator running except where noted
Coleman	Brushless models, except PM500 and 600	Remove the spark plug; make up an extension cord with alligator clips on one end and a 120 VAC plug on the other. Observing the correct battery/receptacle polarity, connect the extension cord to a 12-V battery, and spin the engine over a dozen times or so. Disconnect the extension cord, and test the genset for voltage output.
Coleman	Brushed	Momentarily connect the lead from the positive terminal of a 12-V battery to the positive brush (which is on the right) and the negative lead to the negative brush or to a good generator ground.
Dayton	1, 2W, 3W series brushed	Same as Coleman brushed.
Deere	225, 325, 375, 500, 650 brushed	Same as Coleman brushed.
Generac	G1000, brushless	Use a 12-V battery to flash the capacitor terminals.
Generac	G1600–G4000, other brushed models	Momentarily connect the positive terminal of a 6-V battery to the positive brush and the negative terminal to the negative brush or to a good generator ground.
Gillette	2.2–15- kW models, brushless	Connect a 12-V battery to the capacitor for one second (Fig. 3-11).
Homelite	Brushed LR, LRI, and LRX	Momentarily connect the positive terminal of a 12-V battery to the positive brush (which is on the right) and the negative terminal to the negative brush or to a good generator ground.
Marathon	Pancake four-pole	Shut down the generator. Disconnect exciter leads F+ and F– from the AVR. Momentarily connect a 12-V battery to the exciter with the positive lead to F+ and the negative lead to F1. Note that failure to isolate the AVR can damage it. Reconnect exciter leads, and start the generator.
Stamford	BC with AVR, brushless	Shut down the generator. Connect the negative lead from a 12-V battery to the AVR F1 terminal and the positive battery lead through a 1-A, 1000-V diode to the F2 terminal. Failure to include a diode in series with the positive battery lead will destroy the AVR.
Yamaha	EF1200, EF1400, brushless	Momentarily connect a 12-V battery, positive to positive, negative to negative to the capacitor.

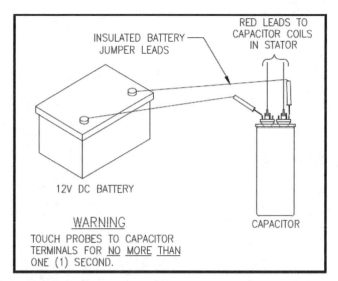

FIG. 3-11   *The Gillette flashing procedure uses the capacitor to introduce battery voltage to the stator. Battery polarity is of no consequence. Storage batteries can release 1000 A or more when shorted. Be careful not to bridge the battery terminals with tools or metal items, and remove all jewelry.*

or hardened rubber mounts, or a loose rotor-shaft bearing, is causing the GFCI to trip.

GFCIs manufactured before mid-2006 require a special tester, available from big-box stores. When newer GFCIs fail, they either go dead or indicate their state of health by shutting off power when the test button is pressed.

Receptacles   Disconnect the wiring to each receptacle under test, insert a jumper wire between the hot and neutral spade terminals, and check for continuity between silver- and gold-colored terminal screws on the sides of the unit.

Voltage-selector switch   The output of dual-voltage generators depends on the way the two main stator windings are connected (Fig. 3-12). When connected in parallel, the windings energize all receptacles at 120 V; connected in series, the output is 120 V hot to neutral and 240 V hot to hot. Figure 3-12 illustrates the circuitry.

Make a careful drawing, with wiring color codes indicated and terminals numbered, before disconnecting the switch. Test for continuity between input and output terminals in both switch positions. All output terminals should receive power from one or another of the active (or wired-up) input terminals.

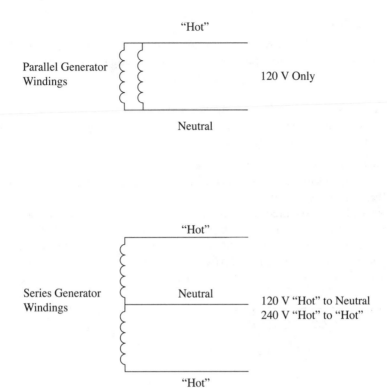

FIG. 3-12 *With windings in parallel, all receptacles see 120 V; connecting the windings in series adds a second 120-V hot lead for a 240-V potential.*

Stator and rotor At least in the United States, rotors are sacrificial items, replaced when they short out or open. Gillette stators can be returned to the factory for rewinding, but owners of other brands are out of luck. The cost of replacement stators, often special-order items with long waiting periods, begins at $300 for the smaller units. The money might be better spent toward the purchase of a new or used generator.

At the minimum, the stator has two main windings and an excitation winding. Additional windings provide power and sensor voltage for the AVR, 12-V battery charging, and engine accessories. Generac 12.5- and 15-kW stators are among the most complex with 12 output leads.

The odor of burnt insulation, carbon smears, or blackened varnish are hard-to-miss indications of stator failure. A massive short to ground causes the engine to labor and bog. Torn insulation suggests that the outboard rotor bearing no longer keeps the rotor concentric with the stator.

Ohmmeter tests

Figure 3-13 illustrates how an ohmmeter placed on its lowest (R × 1) scale is used to test for shorts and opens on Gillette stators and rotors. The same basic procedure is followed for other makes and models.

Stator   A two-pole (3600-rpm) genset has two main windings, each with four leads. A four-pole (1800-rpm) genset has four main four-lead windings. Follow this procedure:

1. Disconnect all lead wires.
2. Place one ohmmeter lead on one of the winding wires and the other lead on its companion wire.
3. Record the resistance value, and repeat the test on the two remaining wires from that winding. Each pair of winding leads should have continuity and the same resistance as its companion pair (#1 and #2 and #3 and #4 in the drawing). Compare your resistance reading with factory specifications if that data are available.

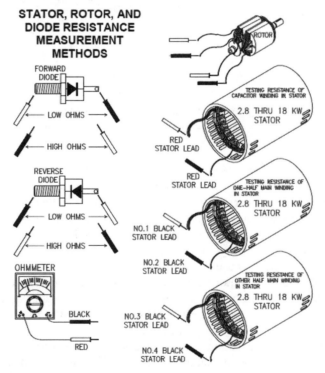

FIG. 3-13   *Gillette suggests that each rotor winding should be tested separately for continuity and shorting to ground.*

4. Also test for grounded field coils by touching one ohmmeter probe to a winding lead and the other to a clean, rust-free ground such as the coil-mounting bracket. Resistance to ground should be infinite.
5. Repeat resistance and ground-fault tests for the exciter winding present on brushless generators.

Rotor The rotor has two windings, each with at least one diode in series on brushless models.

1. Carefully unsolder the diodes from their leads.
2. With an ohmmeter set on the R × 1 scale, measure the resistances of each winding, which should be equal and reasonably close to factory specifications. Check for ground faults between the windings and adjacent rotor laminations.
3. Using an approved electrical cleaner, remove all traces of dirt, oil, or other contaminants from the rotor. Seal the wiring with the appropriate varnish.
4. Test and replace the diodes as described in the "Diodes" section later in this chapter.

Megaohmmeter tests

Although few DIY mechanics have access to a megaohmmeter (megger), some description of the use of these instruments seems appropriate (Fig. 3-14). A megger stresses stator and rotor insulation to detect present or incipient failures that ordinary ohmmeters cannot detect.

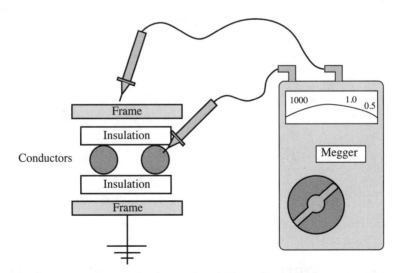

FIG. 3-14 *A megger test is the electrical equivalent of a medical stress test.*

Because of the high voltage, all electrical components must be disconnected and rotor diodes shorted with jumper wires before testing. New, professional-grade stator and rotor insulation exhibits 100 MΩ (100 million
ohms) to ground. Less than 1 MΩ means trouble.

Moisture causes most insulation problems. A generator often can be
restored to service by running it without load for 10 minutes or so. If the
problem persists, an electric heater can be placed at a safe distance from the
fan inlet and the generator run for several hours without load. Be careful not
to overheat the windings.

Ohm's law and resistance

Evaluating windings by their resistance with an ohmmeter is a convenient but
imprecise procedure. The resistances, which often amount to less than 1 Ω, can
best be derived with the aid of a 6-V battery connected as shown in Fig. 3-15.

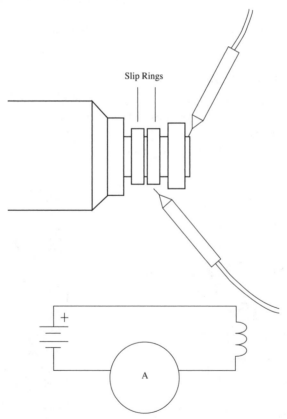

Slip Rings

FIG. 3-15 *A 6-V battery and Ohm's law come to the rescue when measuring very
low values of resistance.*

Suppose that we have 5.70 V at the battery and a current flow of 4.24 A. According to Ohm's law, resistance equals voltage divided by current flow, or amps; that is,

$$R = 5.70 \text{ V} \div 4.24 \text{ A} = 1.34 \ \Omega$$

Note that we cannot measure resistances of less than 0.5 Ω with this method. The rapid rate of battery discharge results in unstable voltage.

Diodes   If the engine is running at rated speed and voltage is only half of what's expected, assume that one or more rotor diodes has failed. The diodes rectify (convert AC to DC) rotor excitation current and act as fuses to open in event of excessive rotor currents.

Test with an ohmmeter set on its lowest scale. Test the diode in the forward position, and reverse the leads to test it in the negative position. As shown in Fig. 3-13, a functional diode should register low, nearly zero resistance in one direction of current flow and high, almost infinite resistance in the other.

Thoughtfully designed rotors have readily detachable diodes, but most have their diodes soldered to their leads. Diodes sometimes test okay and refuse to function under load. In other words, don't hesitate to change out these inexpensive parts. If either has failed, replace the pair or, as the case may be, all four. An increasingly common practice is to combine a varistor, a kind of electrical pop-off valve, in parallel with the diodes (Fig. 3-16).

Carefully make the solder connections using a surgical hemostat or a pair of long-nosed pliers held closed with a rubber band as a heat sink between the diode and the soldering gun. Apply no more heat and solder than necessary to make a low-resistance joint.

Brushes   Replace the brushes when cracked or worn to half their installed length. Failed brushes can fry their plastic brush holders, parts that on some Briggs and Generac machines double as rectifiers. Polish the slip rings with a Scotch-Brite pad, and wipe off all traces of abrasive.

Capacitors   Brushless and a few brushed gensets employ one and sometimes two electrolytic capacitors as a gesture toward voltage regulation (Fig. 3-17). Most are accessible with the generator end cover removed. Generac and several Robin machines have dual capacitors mounted behind the control panel.

WARNING: Capacitors hold a charge after the genset has stopped. Before making any tests, bridge the terminals with a screwdriver with an insulated handle. These large capacitors kick like a mule. A better approach, one that does not risk damage to the dialectic layer, is to discharge capacitors through a moderate load.

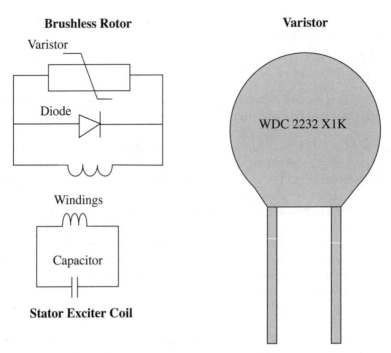

FIG. 3-16 *By exhibiting low resistance to large current flows, a varistor shunts surge currents around the rotor diodes. When tested with an ohmmeter, a varistor will show high resistance in both directions. These components, all of which are identified by an alphanumeric code, can be purchased at electronics supply houses.*

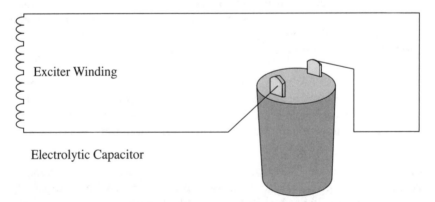

FIG. 3-17 *The odd-shaped electrolytic capacitors used by Honda and Generac can be replaced with any generator capacitor of equivalent size and voltage rating. You may need to secure the new parts with nylon tie wraps.*

Capacitors fail becoming nonconductive or by shorting their plates together. Each type of failure will result in zero generator output. If the top of the capacitor—the terminal end—bulges outward, the device has failed and must be replaced.

To test, disconnect the wiring to the capacitor, and with the engine shut down, connect an analog ohmmeter (one with a needle) to the terminals. Set the meter scale at R × 100. The needle should move to the right, showing a progressive increase in resistance as the capacitor charges. Once at full charge, resistance should remain constant. Fluke VOMs bleep if the capacitor tests okay.

Genset capacitors carry a 450-V rating and range in size from 18 to 100 mF. Replace with one of the same rating and designed for generator service. Starting capacitors for electric motors quickly fail when used in a generator.

Dayton and several other manufacturers use small, metal-encased capacitors to suppress radiofrequency (RF) noise in output circuits (Fig. 3-18). Output will be lost if an RF capacitor shorts to ground. Test with an ohmmeter as described for electrolytic capacitors.

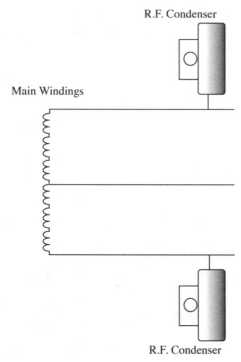

FIG. 3-18 *A few gensets use RF capacitors, similar to those used to dampen voltage arcs across the contact points in old-fashioned ignition systems. This may be why they are called "condensers" rather than "capacitors," which is the modern term.*

Rotor-shaft bearings Noise and voltage instability are early signs of bearing wear as the air gap between the rotor and stator changes. Further wear permits the rotor to come into rubbing contact with the stator, junking both parts. Bearing seizure, or spin, wallows out the bearing boss or, less frequently, scores the rotor shaft.

Most bearings require a gear puller to remove and a driver sized to match the inner race diameter to install (Fig. 3-19). Do not apply force to the outer race. Note the position of the original bearing, which usually beds against a shoulder. The manufacturer's code should be on the outboard side of the bearing since this face is intended to withstand installation forces. Bearing supply houses can provide a replacement, often of better quality than the original.

On Honda EB, EG, and EM gensets, the bearing secures to the end cover with a half-moon retaining ring, an approach that requires special assembly instructions (Fig. 3-20).

A replacement bearing support is an expensive, often difficult-to-find part. If the bearing boss has wallowed, you might try anchoring the new bearing with Loctite 601 Bearing Fit or 638 Maximum Strength Retaining Compound. Home machinists can fabricate a bushing, which is a bit more complex than it sounds because rotor and stator centers must be located precisely. A small misalignment shows up as rotor-to-stator contact when the machine reaches operating temperature.

Rotor extraction The rotor shafts of single-bearing generators normally have a male taper on their inboard ends that matches a female taper on the crankshaft. To separate the rotor, undo the central bolt that secures the assembly. A strap wrench prevents the flywheel from turning as the bolt is loos-

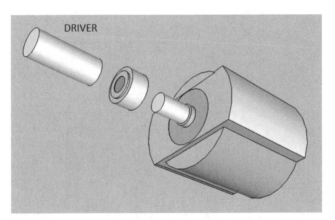

FIG. 3-19 *A suitable driver has an inside diameter (ID) slightly larger than the rotor shaft outside diameter (OD) and an OD just smaller than the OD of the bearing inner race. The working end should be dead flat.*

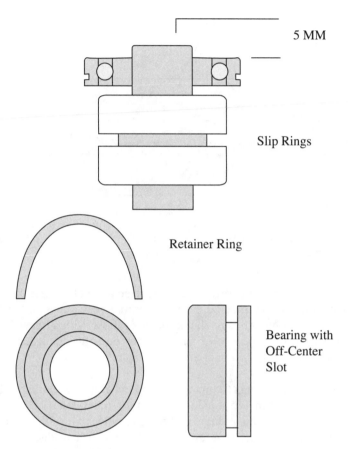

5 MM

Slip Rings

Retainer Ring

Bearing with
Off-Center
Slot

FIG. 3-20 *Honda retainer rings index with grooves in the bearing OD and the generator end cover. Rings install with the raised section in the deeper part of the bearing groove.*

ened. One can also lock the piston with nylon cord fed into the cylinder through the spark plug or injector port.

Determine the diameter and thread pitch (Generac uses American threads) of the hold-down bolt and the counterbore in the rotor shaft. Make up a stud with threads on one end that match the center bolt, slot the other end for screwdriver purchase, and thread it into the crankshaft stub. The free end of the stud should come to within 3/4 in. of the end of the rotor shaft, as shown in Fig. 3-21. The jack bolt threads into the rotor shaft. Tightening the bolt against the end of the stud separates the two shafts.

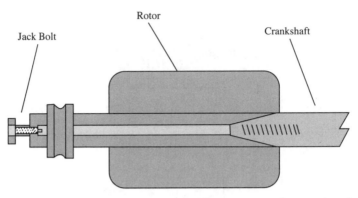

FIG. 3-21 *As shown on the right, most rotors have provision for a stud and jack bolt. If the jack bolt binds, strike it with a hammer, back off a quarter-turn, and retighten. Remove any rust or burrs from the tapers prior to assembly, which is made dry, without lubricants.*

Some early DeVilbiss and other, mostly vintage generators lack threads for a jack bolt. These rotors may be extracted with a gear puller or hardwood wedges. Another approach is to shock the rotor loose with a heavy hammer and a brass drift. Drive the rotor shaft deeper into the crankshaft socket. Hopefully, the taper will spring open enough to release.

Rectifiers   Exciter circuits for many brushed generators include a full-wave rectifier. Briggs and Generac sometimes integrate rectifying diodes with the brush holders; other manufacturers make the rectifier part of the AVR. A full-wave rectifier usually has four diodes connected in a bridge (Fig. 3-22). Kohler, Homelite, and a few other manufacturers achieve full-wave rectification with two diodes and a center-tapped coil.

Full-wave rectifiers are also found on battery-charging and engine-control circuits (Fig. 3-23), although many manufacturers compromise and use a single diode. When this is done, half the AC output is lost during rectification. Test diodes with the VOM on the R × 1 scale, as shown in Figs. 3-24 and 3-25.

Several Honda full-wave rectifiers have only three terminals, which at first sight can be confusing. Check the diodes as described in in the caption to Fig. 3-25.

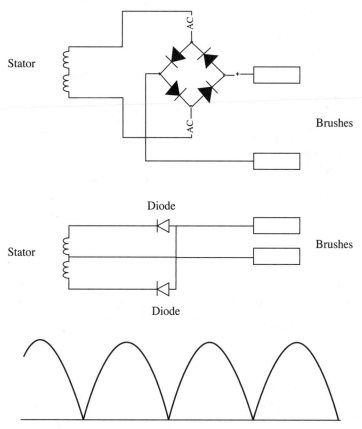

FIG. 3-22 *Full-wave rectifiers are an essential component in many brushed generator excitation circuits. Four diodes or two diodes feeding from a center-tapped coil produce the unidirectional waveform needed for rotor excitation.*

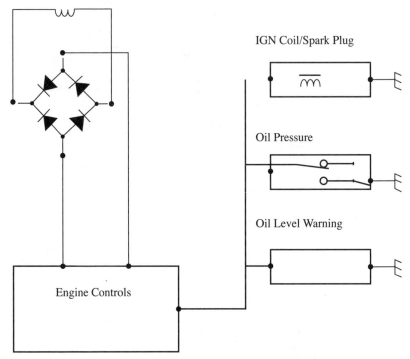

FIG. 3-23 *Many gensets employ a rectifier and an auxiliary winding (sometimes under the flywheel) to provide DC for battery charging and engine accessories.*

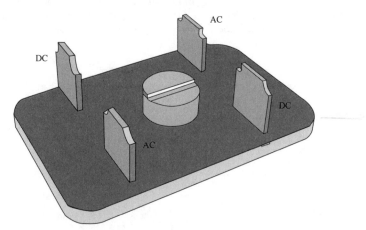

FIG. 3-24 *A four-terminal bridge rectifier. Diagonal terminals should conduct test current in one direction and exhibit high resistance in the other direction.*

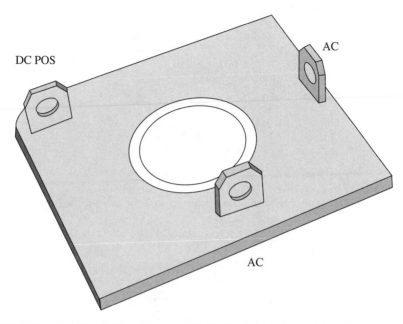

DC POS

AC

AC

FIG. 3-25 *Test Honda three-terminal diodes by placing the positive ohmmeter probe on the positive terminal and touching either of the AC terminals with the negative probe. Reverse the meter leads. There should be conductivity in one direction and nearly infinite resistance in the other. Repeat the test for the remaining AC terminal. Note that the DC terminal on many of these units is adjacent to the beveled edge of the rectifier case.*

Transformer A transformer consists of a primary, or input, winding and a secondary, or output, winding over a laminated iron core (Fig. 3-26). Test with an ohmmeter to determine that the windings are insulated from each other and the iron core. The resistance imposed by the windings should be quite low, almost undetectable with a VOM.

AVR The AVR mounts at the outboard end of the generator head or behind the control panel. Appearance varies—some are encased in plastic or aluminum, and others consist of discrete components on an open circuit board (Fig. 3-27).

Nearly all these units have a voltage trimmer and a second adjustment for frequency. Industrial-quality AVRs have additional adjustments.

Adjust for voltage and frequency as follows:

• Start the generator, and verify that it turns 3750 rpm (±50 rpm).
• Adjust the voltage trimmer to produce 125 to 130 V.

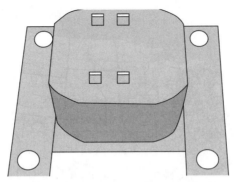

FIG. 3-26 *Transformers may be present on excitation, engine-control, and output circuits. Test each pair of terminals for continuity and for shorts to the frame.*

(a)

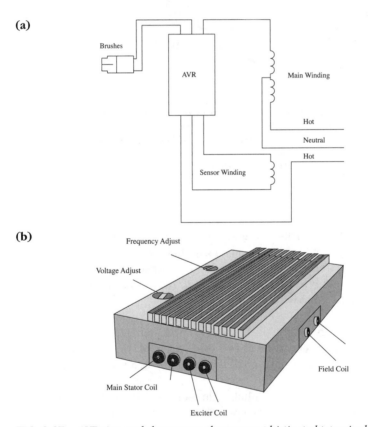

(b)

FIG. 3-27 *AVR size and shape vary; the more sophisticated types include cooling fans and at least one adjustment screw for voltage. The schematic (a) illustrates an AVR used with a brushed genset. A main stator winding (b) provides power for the unit, and the sensor winding reports output voltage.*

- Adjust the frequency trimmer for 61 or 62 Hz. If you're working without a frequency meter, connect a 60-W incandescent bulb across the output, and adjust as needed to eliminate flicker.

Diagnosing an AVR problem often comes down to substitution. Even with the factory data, resistance test results often fall into a gray area, somewhere between "pass" and "fail." Honda has no tests for EN3500 and EN5000 AVRs. However, Table 3-3 can narrow the possibilities somewhat, as does the rotor-excitation test described in the following section. At any rate, check and recheck the other components in the excitation circuit before purchasing a new AVR, which can cost more than $200. Vendors do not make refunds on electrical parts.

Table 3-3
AVR troubleshooting

Fault	Possible cause	Corrective action
No output voltage	Disconnected wires to AVR	Check wiring
	Rectifier bridge failure	Replace rectifier
	Defective AVR	Apply excitation current and replace AVR as necessary
Low output voltage	Insufficient generator rpm	Check engine
	Disconnected wires to AVR	Check wiring
	Defective diode in bridge rectifier	Replace rectifier
High output voltage	Disconnected wires to AVR	Check wiring
	AVR voltage setting incorrect	Adjust voltage trimmer
	Defective AVR	Replace AVR
Unstable output voltage	Engine speed does not remain constant	Correct problem with engine or governor
	AVR stability setting incorrect	Adjust stability trimmer
	AVR wiring connections incorrect	Correct wiring
	Defective AVR	Replace AVR
	Fault in generator output circuitry	Identify and correct problem
Voltage drops under load	Fault in generator output circuitry	Identify and correct problem
	Defective AVR	Replace AVR

Rotor excitation test

An infusion of battery voltage to the rotor can be very helpful during diagnosis. If energizing the rotor has no effect on output, the range of malfunctions is reduced to the rotor windings, diodes (when present), and components in the output circuit. On the other hand, if battery voltage does produce output, the problem must lie in the exciter circuit.

This test works best for brushed gensets that limit application of solid-state components to the AVR. It is not recommended for the newer, more complex machines with printed circuit boards and integrated control circuitry.

Using a fully charged 12-V battery, make the test as follows:

- Disconnect the AVR from the exciter circuit. If the genset doesn't have a voltmeter, set your VOM on the 250 VAC range, and insert the probes into a 120-V receptacle.
- Start the generator, allow it time to warm up, and verify that it turns a steady no-load 3750 rpm (±50 rpm).
- Connect one lead from the battery to the rotor brush with the matching polarity, which should be marked on the brush holder.
- Leave one battery lead disconnected.
- Flick the loose battery lead into contact with the other brush while watching the voltmeter. A second or less of contact should be sufficient to get a voltage reading that, depending on the genset, will range from a low of 50 V or so to more than 150 V.

If the transfusion brings a dead genset back to life, you can be confident that downstream components—receptacles, GFCI, breaker, and associated wiring—function. The exciter circuit is at fault.

Inverter generators

Conventional generators resemble the stereotype of Italian drivers: they have two speeds—wide-open throttle and off. Actually, most of these generators do shed rpm when under light loads, but the correlation between generator rpm and load is crude.

Inverter generators uncouple rpm from voltage and frequency. Consequently, these generators spend much of the time at part throttle, where they consume less fuel, make less noise, and emit fewer pollutants. Output from the better examples equals or exceeds the standards for commercial power. Consequently, computers do not overheat, sound systems retain fidelity, and battery charging proceeds quickly. In addition, inverter generators weigh 25 or 30 percent less than conventional types, which is an important feature for smaller, hand-carried units.

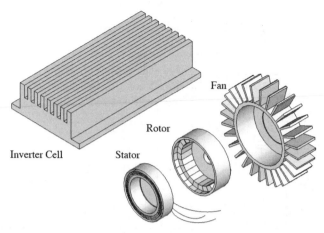

FIG. 3-28　*Key elements of an inverter generator are a multipole stator, a permanent magnet rotor, an inverter cell, and an oversized fan. The rotor, which may function as the engine flywheel, cantilevers off the crankshaft without the need for an outboard bearing.*

Figure 3-28 illustrates the basic components. The generators we have been discussing encase the rotor within the stator. Inverter generators do the opposite: the rotor surrounds the stator, and rather than use power-consuming windings, the rotor energizes the stator with permanent magnets. Nor is there anything resembling an excitation circuit. A stepper motor matches engine speed to load.

Unfortunately, the technology is expensive and, at present, limited to about 6 kW. Although many components are similar to those used on conventional generators, the inverter cell does not lend itself to repair. And replacements are expensive—a discount supplier lists Honda EU3000ie cells at $540.

Operation

Power output from an inverter generator undergoes three transformations (Fig. 3-29). The generator produces high-frequency three-phase current,

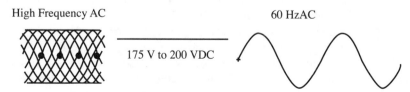

FIG. 3-29　*Output undergoes three transformations from high-frequency AC to DC and finally to 60 Hz AC.*

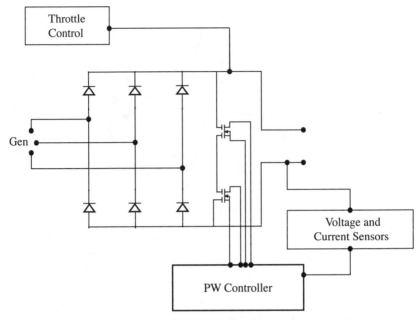

FIG. 3-30 *Two switching transistors are used in this circuit that with more elaboration would produce a square-wave output. The PW controller, responding to feedback from the voltage and current sensors, adjusts the duration of metal-oxide semiconductor field-effect transistor (MOSFET) conductivity to compensate for load. The throttle control drives a stepper motor to match engine speed with load.*

which is rectified into direct current. Solid-state switches then chop the 180 to 200 VDC into segments that are reconstituted into 120 V, 60 Hz AC. This last bit of sleight of hand is accomplished by pulse-width modulation (PMW) (Fig. 3-30).

Most inverter generators use MOSFETs as switches, which, like all switches, do not modulate. A MOSFET has only two states—on and off. But AC undulates, smoothly rising to a peak, falling to zero, and repeating the pattern in the reverse direction. The usefulness of an inverter generator depends in great measure on how well its ON-OFF switches mimic the 60-Hz sine wave.

Microprocessor controllers stagger MOSFET ON-OFF times with reference to a 120-VAC 60-Hz signal. Figure 3-31 shows how pulse width—the ON time—changes to conform to the sine-wave profile. Pulses are narrow as the wave rises, flatten at the peak, and narrow again as zero is approached. But MOSFETs and integrated-circuit controllers are expensive. Tailgate inverters, devices used for camping and rural get-togethers, are suitable for incandescent lighting and resistance heaters (Fig. 3-32). Moving up in sophistication and cost, most inverter generators have a two-step output in the form of what is somewhat

Pulse Width

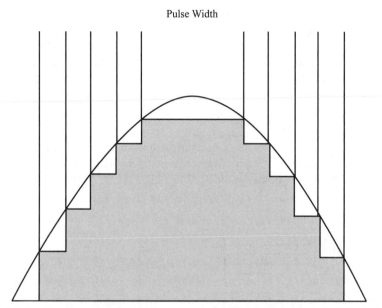

FIG. 3-31 *Microprocessor controllers vary the width, or duration, of MOSFET ON pulses to yield a sawtoothed facsimile of a sine wave.*

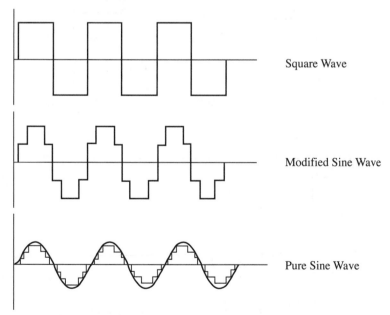

Square Wave

Modified Sine Wave

Pure Sine Wave

FIG. 3-32 *The quality of the sine wave accounts for much of the price differential among inverter generators.*

optimistically called a *modified sine wave*. Although subject to harmonics, this waveform is compatible with most appliances and construction tools. The best inverters incorporate more steps in sine-wave generation to yield an output that can power computers, uninterruptable power supplies, and the digital controls used on Energy Saver washing machines and other upscale home appliances. Each step requires a pair of MOSFETs and a controller.

Troubleshooting

Figures 3-33 and 3-34 show inverter connections for the Honda EU3000ie and an inexpensive import. Other inverter generators have similar layouts: the inverter is the wiring nexus with:

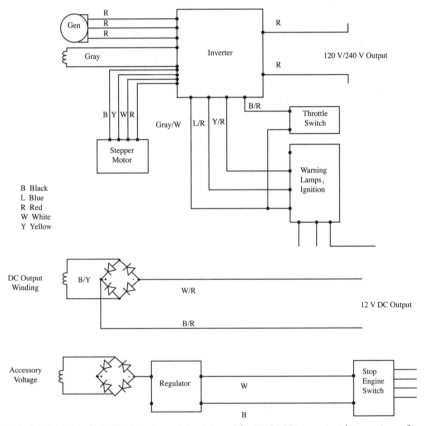

FIG. 3-33 *Honda EU3000ie inverter wiring. The EU2000i is a simpler version of this machine that shares the basic layout and uses the same color codes.*

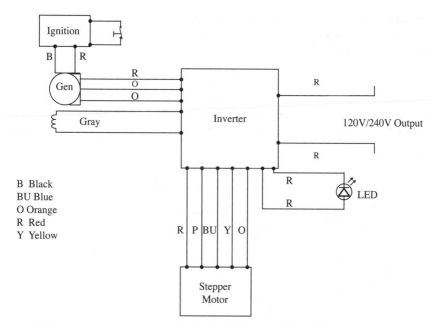

FIG. 3-34 *The Duracell 10R1i inverter generator develops 1200 W of surge power and 1000 W of rated power.*

- Three leads going to it from the main stator windings
- Two signal leads from an auxiliary stator winding
- Four or five leads from the throttle-control stepper motor
- Two output leads
- Leads from the control panel (Honda inverters also supply power to a throttle switch)

Begin by measuring no-load output voltage at the receptacles with the engine turning at full rpm. Because engine speed has no bearing on frequency or voltage, inverter generators have no-load speeds in excess of 3600 rpm. Output should range between 115 and 128 V, give or take a few volts.

If receptacle voltage is below specification, stop the generator and unplug the stator connection at the inverter. The three wires from the main windings will be distinguished by a common color code. They should exhibit the same resistance and be isolated from ground. If low resistance values make your VOM unstable, try measuring resistance indirectly as a function of amperage, as described under "Ohm's law and resistance."

To further verify stator functioning, you may want to measure stator output voltage. This operation requires making the same series of measurements described earlier with the generator running and the VOM set on its 0- to 500-VAC scale. Expect to see something on the order of 200 V from each of the three S lead pairs.

WARNING: These voltages are quite lethal.

A functioning stator and no or low output at the receptacles suggests that the inverter is at fault. Before condemning the unit, though, make a resistance test of the exciter winding (*E* in Fig. 3-35), and verify that panel connections

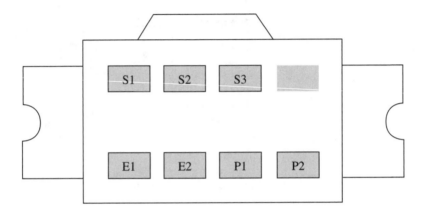

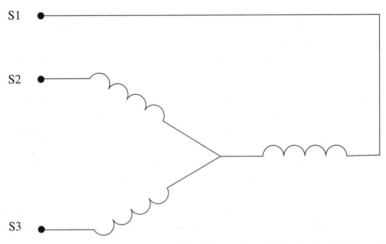

FIG. 3-35 *Inverter connectors vary with make and model but include at least five inputs—three stator main windings (S) and two from the exciter winding (E). Two or more output leads go to the control panel (P). S lead-resistance values should be identical.*

are sound and that the main breaker, GFCI, and other components in the output circuit function normally.

Problems with the 12-VDC circuit usually involve the bridge rectifier. Test as described earlier under "Rectifiers." If the rectifier is okay, test the 12-V stator winding for continuity and shorts to ground.

4

Engine electrical

This chapter describes how to troubleshoot and repair engine electrical components.

Engine management controls

Engine management controls adjust rpm to load, enrich the air-fuel mixture during cold starts, and shut the engine down when temperatures are excessive or oil pressure inadequate.

Power sources

Figure 4-1 shows a very common arrangement, consisting of a 12-V stator winding, a full-wave rectifier, and a battery. The engine block and frame function as the ground return. Some of the charging circuits are fused to protect the winding from the high current flows encountered when charging a dead battery. However, battery–charging circuits have little by way of filtration or voltage regulation, and should not be used to power digital equipment.

Smaller gensets power engine controls and sometimes a battery from a flywheel alternator. Permanent magnets cast into the flywheel rim energize one or more pairs of stator windings to produce AC voltage. Output is intermittent because most under-flywheel alternators have only a single pair of windings (Fig. 4-2).

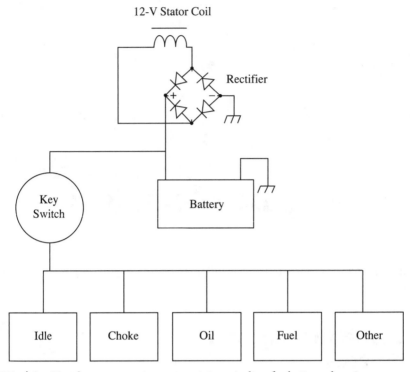

FIG. 4-1 *Most larger generators use a stator winding for battery charging.*

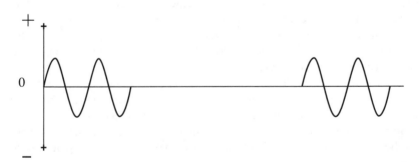

FIG. 4-2 *Output from a two-coil flywheel alternator. Output goes flat until the magnets again swing into position adjacent to the coils. Multiple-coil alternators have much smoother outputs.*

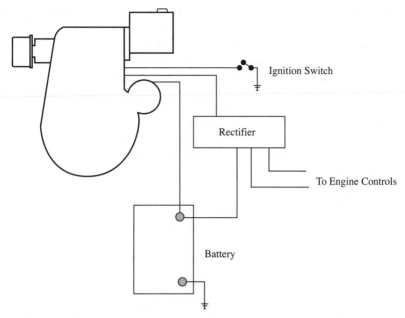

FIG. 4-3 *The flywheel alternator on smaller gensets supplies power for battery charging and engine controls.*

A half- or full-wave rectifier (which may be integrated with a voltage regulator) converts the AC voltage from the under-flywheel winding to DC (Fig. 4-3). With the engine turning at 3600 rpm, you should see at least 13.5 V at the battery terminals. If not, verify that the connectors make good contact, and perform resistance tests on the rectifier as described in Chap. 3. If resistances show the normal low and infinite pattern as the ohmmeter probes are reversed, start the engine, and verify that the unit receives AC power. Single-diode rectifiers may appear as a bulge in the wiring. Bridge rectifiers sometimes are hidden under the blower housing.

Fuel cutoff valves

There are two types of fuel cutoff valves, each with a distinct purpose. All gensets worthy of the name have an in-line shutoff valve, usually manually operated. Without such a valve, the carburetor bowl would be constantly replenished as the fuel in it evaporates, leaving heavy hydrocarbons behind as "varnish." Even if reliability were not a concern, U.S. Environmental Protection Agency (EPA) and California Air Resources Board (CARB) limits on evaporative emissions from small engines have the effect of making fuel shut-off valves mandatory. User-friendly gensets integrate the valve with the key switch. Diesel engines use a fuel cutoff valve (often integrated with the high-pressure pump) as a means of stopping the engine.

In addition to an in-line shutoff valve, emissions-compliant engines have a second, solenoid-operated valve at the carburetor (Fig. 4-4). This normally closed valve denies fuel to the main jet to prevent backfiring that would otherwise accompany shutdown. Lean-burn engines run hot enough to ignite fuel collected in the inlet and exhaust tracks without the benefit of a spark.

An occasional backfire is normal, although the frequency can be reduced by slowing the engine before shutting it down. If backfiring is persistent, suspect that the valve has stuck open. (A stuck-closed valve would of course prevent the engine from running.)

To service the valve:

Step 1. You should feel the valve click as the ignition switch on battery-equipped gensets is cycled on and off. This test establishes that the

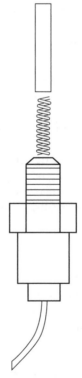

FIG. 4-4 *A solenoid-operated shutoff valve mounts on the carburetor to prevent backfires during engine shutdown.*

valve receives power and that the solenoid functions. However, it is still possible for the plunger to stick closed.

Step 2. Shut off the manual fuel valve, disconnect the lead to the solenoid, and remove the valve assembly from the carburetor. Wipe up the fuel that spills and allow time for the residue to evaporate before continuing.

Step 3. Work the plunger by hand to verify that it moves freely. Should it bind, clean the parts with lacquer thinner or a carburetor solvent.

Step 4. You may want to verify valve function by applying battery power to the solenoid. Some solenoids have an external ground wire; others ground internally to engine metal.

WARNING: This last test, which will be accompanied by sparking, should be made on the bench. If you use the genset battery as the power source, make certain that no trace of spilled gasoline remains on the machine.

Honda complicates matters on its larger engines by using a normally open solenoid that closes only during the coast-down period after ignition is denied (Fig. 4-5). No solenoid voltage is present at other times. This is accom-

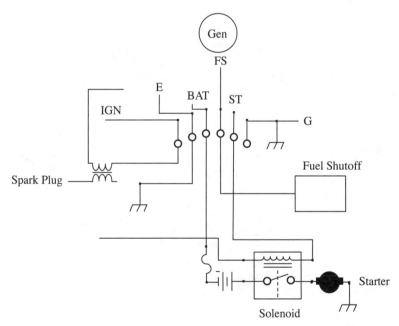

FIG. 4-5 *When turned OFF, the ignition switch used on some Honda generators shorts out the primary ignition coil winding (IGN goes to ground through E) and connects the fuel shutoff valve (FS) to generator power.*

plished by powering the solenoid from the generator (even through the genset may have a battery). When the generator slows to the degree that it no longer produces power, the solenoid opens. Input voltage checks must be made quickly during the brief coast-down window. If no voltage is detected, check the rectifier and, as necessary, work upstream from the rectifier to the associated stator winding.

Idle control

Idle is too strong a word for reducing engine speed by 30 or 40 percent under partial load. But the speed reduction does save fuel and probably adds a few score hours to engine longevity. An electronic control unit, easily recognized by the loop through which sensor leads pass, signals a carburetor-mounted solenoid or stepper motor to reduce the throttle angle under light loads (Fig. 4-6). Most gensets use the main output leads for sensing; others sense current in a dedicated stator coil. Honda sometimes incorporates the sensor winding into the flywheel alternator.

Response to load is not instantaneous. The engine requires some time to overcome rotating inertia and respond to the suddenly opened throttle. Consequently, the idle option should be disengaged when large reactive loads are anticipated.

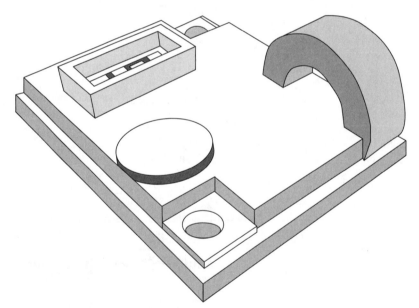

FIG. 4-6   *Idle-control sensors respond to the intensity of the magnetic field surrounding the leads that pass through the powdered-metal loop. Clamp-on ammeters employ a similar technology.*

If the generator fails to idle under light load:

- With the machine shut down, disconnect and make a resistance test of the idle-control switch for continuity and shorts to ground.
- Do the same for the idle solenoid or stepper motor.
- Should these tests reveal that nothing is wrong, start the generator, and measure the output voltage from the idle-control unit at the throttle solenoid or stepper motor. A stepper has multiple leads energized as the control unit dictates. If no voltage is present, check that the control unit receives operating voltage before assuming that it is bad.

Solenoid-controlled throttles have an adjustment to set the idle speed: cheaper units would have you bend the solenoid support bracket, but most enable adjustment by means of a slotted solenoid mount or a threaded solenoid plunger.

To make things interesting, smaller Honda gensets induce idle with a vacuum actuator and solenoid-operated valve (Fig. 4-7). Normally the valve bleeds a small amount of engine vacuum to the atmosphere. But under light loads, the control unit signals the valve to close, diverting vacuum to the actuator, which responds by reducing the throttle angle.

Disconnect the vacuum line to the actuator, and depress the plunger while sealing the vacuum port with a finger. If the diaphragm is sound, the plunger will remain retracted against spring tension. A more precise test requires a hand-operated vacuum pump of the sort sold at auto parts stores.

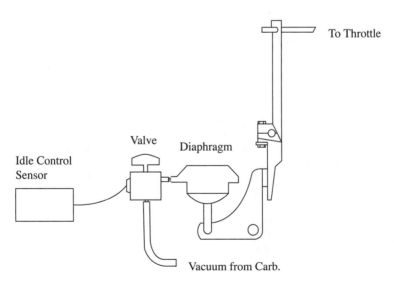

FIG. 4-7 *Honda uses a three-way vacuum valve and actuator to slow the engine under reduced loads.*

Automatic choke

Automatic-choke malfunctions are fairly common. A stuck-open choke can make cold starting difficult or impossible. A choke that remains fully closed after operating temperature is reached richens the mixture enough to shut the engine down. A partially closed choke wastes fuel, blackens the exhaust, and costs power.

If you're experiencing choke problems, first verify that the carburetor butterfly moves easily on its pivots. If necessary, clean the bearing surfaces with lacquer thinner or an aerosol carburetor cleaner. But do not attempt to remove the butterfly—the screws that secure the butterfly to its axle will almost surely strip out or shear off.

Older generators often have a Delco-style choke, familiar to readers old enough to have worked on carbureted automobiles. The choke actuating mechanism consists of a bimetallic coil spring surrounded by a 12-V heater element. As the spring is warmed, it expands to generate force that opens the choke butterfly. Adjust spring tension so that the butterfly remains partially open on a cold engine. The specification varies, but 1/4 to 3/8 in. of clearance between the upper edge of the butterfly and the carburetor air horn should get the engine started. Some experimentation will be required to find the ideal setting.

Homelite and Mitsubishi use a flat bimetallic spring that in the case of Mitsubishi is supplemented by a vacuum actuator (Fig. 4-8). If the choke is

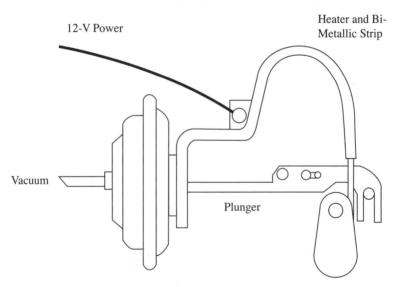

FIG. 4-8   *The bimetallic spring consists of two metals, bonded back to back, with different rates of thermal expansion. As the spring heats, it expands to straighten itself and open the choke.*

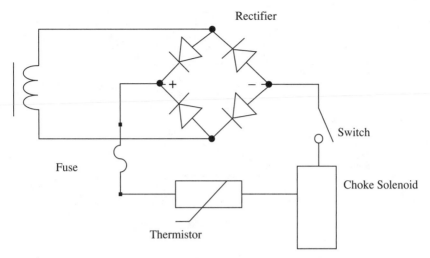

FIG. 4-9 *A thermistor-controlled solenoid.*

slow to open, inspect the butterfly for freedom of movement, test the vacuum actuator with your finger as described earlier, and inspect the vacuum line for cracks.

Several Honda models activate the choke butterfly with a solenoid (Fig. 4-9). A positive-temperature-coefficient (PTC) thermistor—a resistor that undergoes a radical increase in resistance as it heats—mounted on the cylinder head cuts off power to the solenoid when the engine warms. Honda thermistors, which the company calls *thermoswitches*, have model-specific trigger temperatures of between 85 and 100°F (29 and 38°C).

Test with an ohmmeter. If the thermoswitch shows infinite or zero resistance, it or the associated wiring is bad. You may also want to apply a small amount of heat to the switch, which should register a sharp increase in resistance as the trigger temperature is reached.

Other thermistors have a negative temperature coefficient (NTC), which means that resistance decreases with heat. NTC thermistors, wired to the primary side of the ignition, act as sentinels to shut the engine down if it or the generator overheats.

Some Honda EB Series gensets use an auxiliary choke operated by a vacuum actuator (Fig. 4-10). The choke remains closed during cranking to encourage fuel flow through the instrument. Once the engine starts, manifold vacuum pulls the choke butterfly open. Test the diaphragm as described earlier under "Idle control," and examine the vacuum lines for cracks. The check valve—actually a check valve with controlled leakage—should pass air easily toward the vacuum inlet and with some resistance in the other direction.

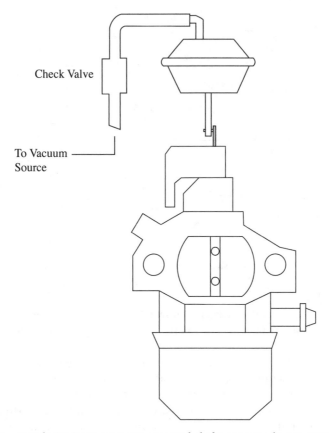

Check Valve

To Vacuum
Source

FIG. 4-10 *Honda EB Series vacuum-operated choke engages during cranking to supplement the manual choke.*

Oil-pressure sensor

The oil-pressure sensor, located near the filter and/or oil pump where it taps the main gallery, is one of the landmarks used to find one's way around genset wiring. This sensor always makes connection with the primary, or low-voltage, side of the ignition. All manufacturers make certain that the engine shuts down once oil pressure is lost.

Most of these sensors take the form of pressure-actuated ON/OFF switches. And most, but not all, are normally open and close when pressure drops below what the manufacturer considers critical. A cutoff pressure of 1 kg/cm² (14.2 psi) is often used.

The sensor responds to low oil pressure by shutting down the ignition and, for many applications, by illuminating a warning light on the control panel. The first priority is to determine whether the sensor is telling the truth. Disconnect the sensor wiring and replace the unit with an oil gauge that has the correct (usually metric) fitting. A far less precise method of detecting pressure is to run the engine for a few seconds and remove the overhead valve cover. The valves should drip with oil.

Oil-level sensor

Like the oil-pressure sensor, the oil-level sensor has veto power over the ignition. And like the oil-pressure sensor, these devices frequently malfunction to shut the engine down within seconds or minutes of starting. To verify sensor function:

- Position the genset on level ground.
- Verify that the oil level is up to the "Full" mark on the dipstick.
- Disconnect the level sensor.
- Start the engine. If the engine continues to run, the sensor is at fault.

Oil foaming can also lead to shutdowns. Suspect this is the cause if the trouble develops immediately after an oil change with a different brand of oil. Level sensors are expensive to replace, and the temptation is to disconnect a faulty unit and depend on the dipstick. But this means that the genset will have to be taken offline every few hours to check the oil level.

Solid-state ignition

An ignition coil is a step-up transformer, similar to those used elsewhere in gensets. It has two windings, a primary and a secondary, wrapped over a laminated-iron armature. The difference between ignition coils and ordinary transformers is in the way the primary and secondary circuits are wired. An ignition coil has one end of its primary winding grounded to engine metal and the other end connected to a mechanical or solid-state switch that, when closed, goes to ground. So long as the switch remains closed, the primary circuit is complete and supports current flow. One end of the secondary also grounds to the engine; the other end makes up to the spark plug, which provides ground for voltages high enough to arc across the spark gap.

Ignition coils have the remarkable ability to step up primary voltage by a factor of 100. As one might suppose, the secondary has far more windings than the primary. For example, the Briggs & Stratton Magnetron has 59.5 turns of wire in the secondary for each turn of wire in the primary. But the main reason for the voltage boost lies elsewhere.

Initially the primary circuit completes itself through the switch to ground. As the flywheel turns, the field produced by the rim magnets induces current flow in the primary, which produces a second magnetic field around primary and secondary windings.

As the flywheel continues to turn the piston approaches its upper stroke limit, or top dead center. The switch that controls primary current opens, breaking the ground connection. The magnetic field around the primary winding collapses upon itself at near light speed. This rapidly shrinking magnetic field induces high voltage in the secondary, which finds ground through the spark plug. Thus, an ordinary little engine, chug-chugging along, generates magnetic field velocities on the order of 300,000 meters per second.

The evolution of ignition systems has focused on the switching element. In times past, contact points functioned as the switch. Today almost all small engines use far more reliable solid-state switches. The Magnetron, illustrated in Fig. 4-11, uses paired transistors, known after their inventor as Darlington transistors, to turn primary current on and off. A trigger coil generates a small voltage that, when present, signals the Darlington transistor to become conductive and complete the primary circuit to ground. When flywheel move-

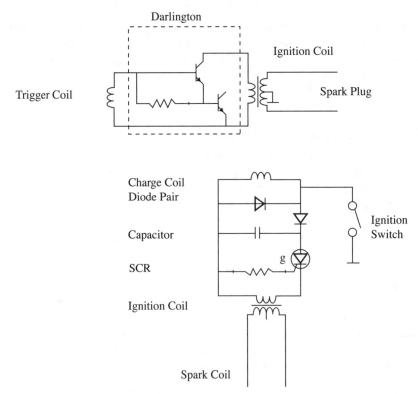

FIG. 4-11 *(Above) The Magnetron has the virtue of simplicity, which may explain why these units almost never fail. (Below) Capacitive-discharge ignitions (CDIs) have the advantage of a rapid rise time, a feature that assists starting and helps to clear fouled spark plugs.*

ment no longer energizes the trigger coil, the Darlington opens, the magnetic field around the primary collapses and the spark plug fires.

A Magnetron also provides ignition advance proportional to engine rpm. At low speeds, the flywheel magnet must come into close proximity to the trigger coil in order to generate the 1.2 V needed to persuade the Darlington transistor to conduct. At higher speeds, less proximity is required, and ignition occurs earlier.

Magnetrons, introduced in 1982, can date the engine. The date of manufacture stamped on the module should be no more than a month earlier than the engine build date. Figure 4-11 also illustrates the circuitry for a CDI system, of the sort popular with Japanese manufacturers. Unlike the Magnetron, CDIs work by inducing rather than collapsing the magnetic field around the secondary-coil windings.

A large capacitor acts as a holding tank to store current induced in the primary winding. Diodes function as check valves to maintain the charge on the capacitor until voltage generated by the timing, or signal, coil turns the silicon-controlled rectifier (SCR) on. The capacitor then discharges, sending a burst of current into the primary windings. This influx creates a powerful magnetic field that induces ignition voltage in the secondary.

Magnetrons, CDIs, and other solid-state systems are self-contained in that they generate primary current with permanent flywheel magnets. Generac, Robin, and some vintage gensets supply primary ignition voltage from the battery.

Flywheel-powered ignitions have a shutdown switch that grounds the primary circuit to the generator frame. The grounding wire sometimes shorts out, causing mechanics to make the expensive mistake of replacing the ignition module. The kill switch for battery-operated ignitions shuts down the engine either by grounding the primary or by disconnecting battery power.

Ignition myths

Briggs & Stratton has done us all a service by pointing out the following ignition-system myths:

- Healthy ignition systems produce blue sparks. *False.* Ignition spark color has no relationship to spark voltage. Blue sparks are no better than the orange sparks produced by Magnetrons. Most spark energy is in the ultraviolet range, which we cannot see.
- Rust on flywheel magnets makes for hard starting. *False.* Rust does not affect a magnetic field. If it did, the Earth would not be a giant magnet.
- Grounding the spark plug against the block and cranking the engine is a reliable method of testing ignition output. *Not necessarily true.* A weak coil can produce enough voltage to fire a spark plug in open air and fail to fire the plug under cylinder compression.

- A wide armature air gap will prevent the engine from starting. *False.* Briggs engineers assert that the air gap cannot be made wide enough to inhibit spark production; the effect will be to slightly retard ignition.

Other than replacing a faulty shutdown lead, solid-state ignition modules are not repairable. Or more accurately, one must be an electronics expert of the first magnitude to defeat the epoxy encapsulation and redo the integrated circuitry. But spark plugs are another matter.

Change out the spark plug(s) at the first sign of hard starting. Table 4-1 lists interchange information for popular brands of small-engine plugs. Note that aluminum heads require brightly plated spark plugs. The plating functions as a lubricant.

New spark plugs usually need to have their gaps adjusted by judicious bending of the side (ground) electrode. A "silver dollar" gap tool, available at auto parts stores, makes this adjustment quick and easy. If you do not know the correct setting, most engines work with a 0.030-in. spark gap. Vintage magneto-fired engines seem happier with 0.020-in.

Gap specifications are not chiseled in stone. Ignition output voltage increases with the gap. Assuming that the ignition module can provide the voltage at cranking speed, a wider than specified gap can assist in starting if the spark plug is fouled with fuel or lube oil. By the same token, a narrow spark gap can sometimes get an engine with poor ignition voltage started.

Install the spark plug dry, without lubricant. Tighten by hand until the gasket makes contact with the cylinder head. Then, using a wrench, tighten the plug about one-half or two-thirds turn more.

Table 4-1
Popular small-engine spark plugs

NGK commercial	Denso	Champion	Autolite
CS1	W9LM US	J19-LM	458
CS2	W20M U	CJ8	255
CS3	W20S U	J8C	302
CS4	W14LM U	J17LM	456
CS5	W14M U	CJ14	258
CS5	K16PR U	R12YC	3924

Starter motors

Batteries are a perennial problem with electric-start machines. Readers new to the subject should understand that lead-acid storage batteries contain sulfuric acid and emit highly inflammable hydrogen gas. Battery explosions are not uncommon. Protect your eyes with splash-proof polycarbonate goggles (designated Z-87) and remove rings, wrist watches and any other all metallic jewelry. You also may want to wear rubber or plastic gloves. To further minimize risk, disconnect the negative (black) cable before disconnecting the positive cable. Tuck the negative cable clear of the battery or metal parts. Then disconnect the positive cable. Reconnect the cables in the reverse order, making the positive connection first.

Disconnect and clean the battery terminal down to bright metal. If cleaning the terminals does not help and the engine still cranks slowly or the starter relay clicks like a disapproving grandmother, the odds are that the battery has lost its charge. Test with a digital voltmeter.

Table 4-2
State of Charge as Indicated by Battery Terminal Voltage

State of charge	Voltage at battery terminals (V)
100%	12.7–12.9
75%	12.4–12.5
50%	12.1–12.2
Discharged	12.0 or less

Note: Batteries develop misleadingly high surface voltages that contribute nothing to cranking power. Measure the voltage after engaging the starter once or twice.

Aside from the battery, starter malfunctions have four possible sources: the starter switch, the wiring between the switch and the solenoid, the solenoid, and the motor. We'll begin with the switch and associated wiring. Should the failure be in this area, turning the starter switch ON has no effect. The solenoid will not click in response, and the motor will not turn over.

1. Find the wire running between one of the smaller terminals at the solenoid and the switch (Fig. 4-12).
2. While a helper holds the switch in the ON position, check voltage to ground at the solenoid connection, the switch, and back to the battery.

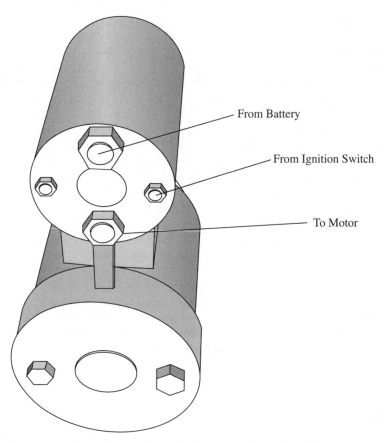

From Battery

From Ignition Switch

To Motor

FIG. 4-12   *The solenoid (actually a relay) may be mounted on the starter motor or remote from it.*

To test the solenoid:

1. Connect a jumper cable to the battery positive (red) post.
2. Touch the other end of the cable to the large solenoid terminal closest to the starter motor. Should the motor respond, the solenoid has failed. If the infusion of current has no effect, the motor has failed.
3. Another, less-civilized way to power the motor is to arc the two large solenoid terminals with a screwdriver. The motor should start amid a shower of sparks.

Starter brushes and shaft bushings are the most likely parts to fail. However, motor parts are extremely hard to find, and you may have to go to a bearing supply house for the bushings and a motor repair shop to find brushes that can be modified to fit.

5

Emergency repairs

Like windshield wipers and parachutes, gensets fail when we need them most. This chapter describes how to identify common malfunctions and make the repairs that, with a bit of luck, will put the generator back into action. Malfunctions that appear suddenly without warning usually can be corrected on the spot.

Not to belabor the point, gensets have a real potential for hurt. Emergency repairs, often attempted in the dark and in inclement weather, do nothing to mitigate the risk. Do not work in the wet or in unventilated spaces. Have a fire extinguisher standing by, and make sure that someone in the household keeps an eye on the operation.

Begin by rereading the operator's manual, which should include a section, however slapdash, on troubleshooting. It is helpful to reconstruct events prior to the breakdown. For example, was the genset running when it failed? Or had it been in storage? Did the problem occur after additional load was plugged in? Has the machine misbehaved in the past? Were air, fuel, and oil filters recently changed? Have the spark plug(s) been changed?

Tools and supplies

Troubleshooting essentials include

- The usual assortment of hand tools, including a volt-ohmmeter (VOM). A tachometer and a frequency meter (such as an inexpensive Kill-A-Watt) are quite useful. You also may want to include gloves and eye protection in the kit.

- Spark plugs.
- Air, fuel, and oil filters. If the engine has accumulated lots of hours, you may want to add a head gasket to the parts inventory.
- Jumper cables to enable an auto battery to be used on electric-start gensets. The cables should be No. 10 American Wire Gauge (AWG) or heavier copper and have copper—not copper-plated—clamps.
- Rags or paper towels.
- Aerosol carburetor cleaner.
- A fully charged fire extinguisher.
- A hand pump to transfer gasoline from an automobile in emergencies.
- A flashlight and spare batteries.

No power

No power from any receptacle

Reset the main breaker on the generator control panel. A breaker that throws as soon as it's set means that the load draws more power than the generator can deliver. Loads, especially reactive loads that occur when motors first come online, may simply be too large. Another possibility is a short to ground in one or more of the loads or in the power cable.

With generator power turned OFF, disconnect the largest load. Switch power back ON. If the breaker trips, progressively disconnect the remaining loads. Should the problem persist, replace the power cord, making sure the generator switched OFF before handling the cord. If the genset has seen extensive reworking or is a new acquisition with an unknown history, low engine speed may be the culprit. Gensets shut off power if no-load rpm drop below 15 to 20 percent less than the rated rpm. Adjust the governor with the aid of a tachometer or a frequency meter. If you have neither and the generator can be persuaded to produce some power, connect an incandescent lamp to the output. Speed up the engine until the lamp no longer flickers.

No power from one receptacle

The same approach works for individual receptacles that have lost power. Reset the breaker or the ground-fault circuit interrupter (GFCI) for the affected receptacle. If the current-limiting device immediately opens, disconnect one load at a time fed by that receptacle, and as a last resort, replace the associated power cord. Make certain generator power is switched OFF before connecting and disconnecting suspect loads and power cords.

Genset slows under load

Test for overloading and short circuits as described previously. However, most of these difficulties are engine-related.

Storage batteries

Batteries are the major problem with electric-start machines. Chapter 4 goes into some detail about the safety precautions that should be taken with lead-acid batteries. I won't repeat those precautions here, other than to remind readers that the negative (black) battery cable should be disconnected before disconnecting the positive (red) cable. Reconnect in the reverse order, leading with the positive cable.

Clean the battery terminals to remove all traces of corrosion. If this does not help, the odds are that the battery has lost its charge. With the engine stopped, the battery should have 12.5 V between the positive and negative terminals. If the meter shows 12.2 V or less, the battery needs a boost.

To jump start a genset from an automobile:

1. Turn the genset ignition switch OFF.
2. Attach one end of the positive (red) cable to the positive terminal of the dead battery. If you're not wearing a face shield, turn away as you make this and subsequent battery connections.
3. Clamp the other end of the positive cable to the positive terminal of the car battery.
4. Connect one end of the negative (black) cable to the negative terminal of the car battery.
5. Connect the other end of the negative cable to an unpainted surface on the genset frame or engine. A connection remote from the battery reduces the risk of explosion. For obvious reasons, do not use the carburetor or fuel line as the negative terminal.
6. Start the car, and let it run for a few minutes at a fast idle.
7. Remove the jumper cables in the reverse order of attachment.
8. Try to start the generator.

Gasoline engine malfunctions

The engine is far more likely to give problems than the generator. In order to start, a gasoline or other spark-ignition engine must have:

- Fuel must be in the right proportion with air. Too much fuel is as bad as too little.
- The spark should have sufficient energy to overcome the effects of compression. That a spark plug fires outside the engine is no absolute, money-back guarantee of ignition voltage.
- Compression, at the most fundamental level, depends on how well the piston rings and the valves seal pressure. Compression also has a dynamic aspect. Unless an engine spins at 250 rpm or so, it does not develop enough compression to start. In other words, the battery should be fully charged or the starter cord pulled with authority.

Preliminaries

Begin by verifying that the fuel is usable. Stale gasoline has a dark orange-brown color and an odor that makes one retch. Water, absorbed from the atmosphere, gives the fuel a cloudy aspect and, in high concentrations, collects as globules that skitter about on the bottom of the fuel tank and float bowl.

Drain the system of suspect gasoline as described later in this section and refill with fresh fuel that can be pirated from an automobile. Renew the paper air-filter element, even if it appears serviceable. The old filter can always be reused.

No start

Go easy on the cranking. If the engine does not start after four or five attempts, further cranking floods the cylinder, runs down the battery, and increases the level of frustration. Waiting an hour for the fuel in a flooded cylinder to evaporate often is enough to get an engine started.

Choke   The choke butterfly must close fully on most carburetors for a cold engine to start. Manual chokes with a direct-acting linkage give few problems. But automatic chokes may not close to the degree necessary or may fail to open fully. A partially closed choke costs power and, depending on the extent of the restriction, can shut the engine down.

Ignition tests   In the absence of some obvious fault, such as fuel dripping from the carburetor or a total lack of resistance to the starter cord, troubleshooting begins with the ignition system. Spark plugs are sacrificial items, replaced as a reflex when anything goes wrong. If changing the spark plug elicits no response, the next order of business is to test ignition voltage. Figure 5-1 shows a homemade spark tester. Briggs & Stratton's PN 19051 features a transparent window over the spark discharge area to prevent ignition of split fuel.

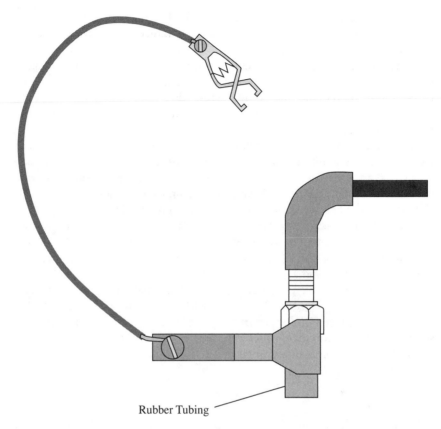

Rubber Tubing

FIG. 5-1 *A homemade spark tester. Rubber tubing over the firing end of the spark plug enables the spark to be seen in bright sunlight. Set the spark gap at 0.060 in. or wider.*

Even when spark is present, it's good practice to repeat the test with the spark plug installed and the tester connected in series. A bad ignition coil can deliver sparks in open air and fail to do so under cylinder compression. You should see a regular splatter of sparks as the starter is engaged. If the system fails to produce a reliable spark, turn to "Solid-state ignition" in Chap. 4 for further information.

Fuel starvation   Once you have replaced the air filter, established that fresh gasoline is in the tank, and checked for spark, the next step is to verify that fuel reaches the combustion chamber. To do this:

- Close the choke or repeatedly press the primer button.
- Crank the engine a half-dozen times.
- Remove the spark plug, which should be damp and smell of gasoline.

If the spark plug tip shows no evidence of fuel, spray a small amount of carburetor cleaner into the combustion chamber. Replace the spark plug, open the choke, and try to start the engine.

If the engine responds to the fuel infusion and continues to run normally, the problem is solved. Engines are sensitive creatures and sometimes need a bit of help to get moving. Should the engine run a few seconds on fuel in the cylinder and die, a stoppage exists somewhere between the tank and the combustion chamber.

As an additional check, remove the air-filter element and spray a little carb cleaner or gasoline into the air intake. If you use gasoine, place the fuel can well clear of the engine. Wipe up any spills, and wait for the residue to evaporate. Install the filter as a backfire barrier, and crank. Once started, the engine should run for a few seconds longer than it did previously.

At this point we know that the engine must have life support in the form of a fuel transfusion to run. The next step is to determine where the supply interruption occurs. This operation entails some fuel spillage, which can be contained with rags and a small can.

There are two basic types of carburetors: those that have a fuel bowl and those that regulate the internal fuel level with a diaphragm (Fig. 5-2). The latter can be recognized by the flat plate under the instrument secured by four screws. These carburetors, widely but not exclusively used on two-stroke engines, are an abomination. Much of Chap. 6 describes how to deal with them.

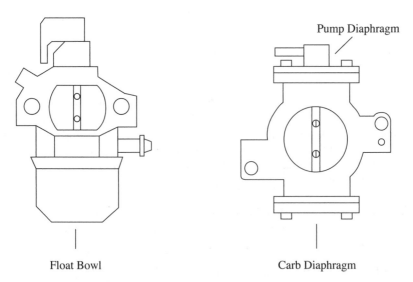

Float Bowl   Carb Diaphragm

FIG. 5-2 *Larger gensets have float-type carburetors. Diaphragm carburetors are pretty well restricted to the two-cycle engines that power hand-carried units.*

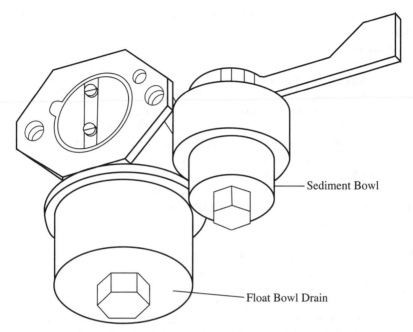

Sediment Bowl

Float Bowl Drain

FIG. 5-3 *Nearly all float bowls have a drain plug. If no plug is fitted, loosen the central hold-down bolt. As shown here, Mikuni and Mikuni carburetors usually come with a sediment bowl.*

Loosen the float-bowl nut or drain plug on the underside of float-type carburetors (Fig. 5-3). If the carburetor is being fed, fuel will run out. Should this be the case, see Chap. 6 for carburetor repairs and cleaning. If the float chamber is dry, the stoppage is somewhere between the tank and carburetor.

For many diaphragm-type carburetors, the availability of fuel can be tested by removing the air-cleaner assembly and pressing the primer bulb. The pump should discharge a stream of raw fuel into the carburetor throat. Another way to verify fuel delivery is to shut off the manual fuel valve and disconnect the fuel line at the carburetor, which can be a bit confusing to identify (Chap. 6 goes into some detail on diaphragm-type carburetor plumbing variations). Fuel should flow from the disconnected line when the manual shutoff valve is opened.

If either type of carburetor does not receive fuel, work upstream from the carburetor, opening each fuel-line connection, until the stoppage is found. Close the manual valve, open the connection, and quickly shut off the flow. Mikuni and Mikuni clone float-type carburetors have an integral sediment bowl and screen assembly that clogs in the presence of stale fuel. The next most likely culprits are the fuel filter and shutoff valve. The last potential restriction is the in-tank filter or screen.

Not all fuel starvation problems are due to stoppages. Air leaks downstream of the carburetor deny fuel to the engine. Make certain that the carburetor mounting bolts are secure. Two-stroke engines complicate matters by using the crankcase as part of the induction tract. Crankcase leaks put the genset out of commission until major repairs can be made.

A restricted exhaust system is another possibility. Owner's manuals suggest that the spark arrestor—a fine-mesh screen placed across the muffler outlet to trap hot carbon particles—should be cleaned every 100 operating hours (Fig. 5-4). Some clog within 50 hours. Remove the screen, and gently clean it with a wire brush. Assemble the fasteners with a metal-based antiseize lubricant such as Permatex 80078. After long service, two-stroke engines also carbon-over their exhaust ports (Fig. 5-5).

Flooding   Severely flooded float-type carburetors dribble gasoline from the air filter. The usual culprit is a particle of rust or sand caught between the inlet needle valve and its seat. Close the fuel shutoff valve, and remove the float bowl, secured by a central nut. Try not to damage the O-ring seal. Briefly open the fuel shutoff valve to wash the dirt out from under the needle. If this does not work, the inlet needle valve must be replaced.

A less dramatic sort of flooding comes about from excessive cranking or from overchoking a warm engine. Liquid fuel collects in the combustion chamber and shorts out the spark plug. Flooding usually can be cleared with a dry spark plug. If the replacement plug fouls, replace it with another. Or simply wait for the excessive fuel to evaporate.

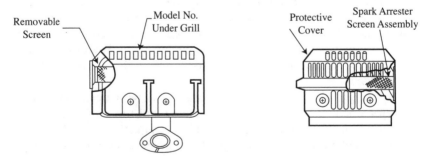

FIG. 5-4   *A spark arrestor consists of a detachable screen placed across the exhaust outlet. The example shown is used on some Honda engines.* USDA

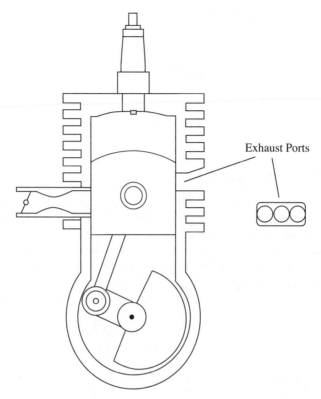

Exhaust Ports

FIG. 5-5 *To clear clogged exhaust ports, lower the piston past the ports, and using a dull tool, scrape off the carbon deposits. Before reinstalling the muffler, pull the engine through several times to clear the cylinder of carbon fragments. Steel mufflers can be cleaned by immersion in hot water and household lye.*

Loss of power as engine warms

This symptom is usually associated with a rich mixture and is accompanied by misfiring, black smoke in the exhaust, and a muted exhaust tone. The spark plug carbons over, often with fluffy deposits that can be wiped off with a finger. The prime suspect is overchoking, either done deliberately by the operator or as a result of a malfunctioning automatic choke. A dirty air filter produces the same symptoms. Plastic foam air-filter elements should be cleaned with detergent and water and should be lightly oiled as described in the owner's manual. Pleated paper filters cannot be cleaned. Water or solvents swell the wood fiber, and compressed air blasts microscopic holes in it.

Sudden shutdowns

Malfunctioning oil-level, oil-pressure, or high-temperature sensors shut the engine down within seconds or minutes of startup. As described in Chap. 4, you can cope with the malfunction by disconnecting the ground lead running from the ignition module to the failed sensor. A few gensets are wired in such a way that disconnecting the lead also disables the stop switch. If this is the case, the manual fuel shutoff valve can be used to stop the engine.

Insufficient power

An ignition tester connected in series with the spark plug will detect misfiring, although a lean mixture—signaled by a bone-white spark plug tip—is the more likely cause of loss of power. Older-model carburetors had mixture adjustments that could compensate for reduced fuel flow. Emissions-compliant carburetors operate with fixed settings that border on lean. A new fuel filter may help, but the usual cure is to dismantle and clean the carburetor. Some owners have found it necessary to replace the carburetor with an adjustable model.

A lean mixture also can be caused by air leaks between the carburetor flange and intake pipe. Check that the carburetor mounting bolts are tight.

It is also possible that the engine lacks sufficient compression, a problem that often can be fixed with a new head gasket, as described in Chap. 7.

Weak compression

Nearly all small four-cycle engines have an automatic compression release, which makes precise measurements of compression impossible. However, if the starter cord exhibits little or no resistance or the electric starter spins the engine without the characteristic whomp-whomp, major repairs are in order.

Diesel engine malfunctions

Diesel engines are much more reliable than their spark-ignition cousins. But compression ignition and the poor suction head developed by injector pumps can present difficulties.

No or slow starting

By far the most common problems presented by well-maintained diesels involve a reluctance to start.

Cold weather Extreme cold reduces the capacity of the storage battery, thickens crankcase oil and increases the amount of compression heat that must be generated to ignite the fuel. One approach that unfortunately only applies to engines with vertical intake manifolds, is to raise the compression ratio with a tiny amount of diesel oil. Remove the air filter and pour a teaspoon of oil into the manifold where it will be inducted into the cylinder during the next intake stroke. Depending upon the size of the engine and its number of cylinders, you may have to add additional spoonfuls of oil. But too much oil hydraulically locks the piston, meaning that you will have to remove the injector(s) to clear the blockage.

Heating the intake manifold with a hair dryer (assuming you have power) reduces the thermal gradient and makes cold-weather starting easier. A propane torch can also be used, but understand that an open flame around greasy engine parts and hydrogen-emitting storage batteries imposes real risks. It's the height of dumbness to end up in the ER because your genset wouldn't start.

All engine manufactures warn against starting fluid. These ether-based aerosols explode violently before the piston reaches top dead center. Most engines, most of the time, tolerate a quick burst of starting fluid applied to the intake manifold during cranking. Spray the whole can of starting fluid into the intake manifold and the engine may very well explode. If you are willing to assume the risk of using starting fluid, be certain that the engine does not have glow plugs. If these incandescent heaters are present, starting fluid ignites as soon as the intake valve opens to create an intake-manifold fire.

The temptation to use starting fluid or an open flame can be difficult to resist, especially at midnight in a rain storm. All of us who work with these engines have at times resorted to these expedients. But readers should recognize the risks involved, both to the equipment and to themselves.

Slow cranking The engine must turn 250 rpm or so to develop enough compression heat for ignition. If the engine appears to crank more slowly than normal, remove the battery cables and clean the cable connectors and battery posts. If the genset has a history of weather exposure, disassemble and clean the battery connection where the negative (black) cable mounts to the engine or genset frame.

Verify that the battery is fully charged. If necessary, connect jumper cables from an automobile as described earlier. Some of the better gensets have two 12-V batteries, connected in series to provide the additional capacity that compression-ignition engines need.

Should a clean cable connection and fully charged battery fail to restore normal starting rpm, assume that the starter motor has wallowed out its bushings. Unless you have a spare starter motor on hand, repairs will have to wait on parts.

Normal ambient temperature and cranking speed   No start with the engine spinning over at its normal rpm almost always involves air in the fuel lines. Air enters whenever a fuel filter is changed, a line opened or the tank run dry. See Chap. 6 for instructions on purging the system.

Runaway

Diesel engines, like nuclear reactors, can run away. This is an extremely rare malfunction that few mechanics have ever witnessed, but it can and does happen. The intake manifold has no throttle valve, a feature that reduces pumping loses and improves fuel economy. How fast the engine turns depends solely upon the amount of fuel it receives. Old, well-worn engines sometimes draw crankcase oil past their rings. The engine accelerates ever faster until the connecting rod breaks. Should you have the misfortune to encounter a runaway, run yourself to get clear of the shrapnel. Brave souls put have shut down these engines by stuffing a rag in the intake manifold to deny air for combustion.

6

Fuel systems

Gensets are especially prone to fuel system malfunctions, probably because these machines spend so much time in storage. But one should not jump to conclusions. The first response to any malfunction is to replace the spark plug. Should that not solve the problem, test for spark and for fuel delivery as described in Chap. 5. If fuel delivery is the culprit, verify that fuel reaches the carburetor before assuming that the carburetor itself is at fault. In other words, work from the assumption that faults have simple, easy-to-fix causes. Most of the time you will be right.

Carburetors are remarkable devices, the fruit of more than a century of development. The carburetor used by the Wright brothers consisted of an exhaust-heated hot plate that boiled off gasoline into a combustible vapor. The pilot controlled engine rpm, more or less, by advancing and retarding the spark.

More sophisticated designs followed. Float valves kept the level of fuel within the carburetor (and, hence the air-fuel mixture) constant. But engines need richer mixtures as loads increase, and until the mid-1920s, these adjustments were the responsibility of the operator. Aircraft imposed other problems. G-forces generated during violent maneuvers slosh the fuel about, upsetting the air-fuel mixture. During World War II, Bendix developed a pulse-operated diaphragm carburetor that opened a fuel inlet valve in concert with engine rpm. Derivatives of those early aircraft carburetors have become universal on small, hand-carried gensets and other portable tools. Larger gensets, intended for horizontal operation, have less temperamental float-type carburetors.

Basic carburetor theory

Carburetors are a less-than-obvious technology: their workings cannot be understood without some grounding in theory. Mechanics who lack this understanding cover their ignorance with new parts and sometimes new carburetors.

Carburetors work by the partial vacuum the engine creates during its intake stroke. The difference between engine-induced negative pressure and atmospheric pressure drives fuel through the jets, which is then atomized into a fine mist in preparation for vaporization. Unlike early carburetors, the ones we deal with have automatic mixture control. During cold starts, the mixture is gasoline-rich, with roughly six or eight parts by weight of air to one part of gasoline vapor. As the engine warms, the mixture leans out, with proportionally more air. At full rated power, the typical air-fuel ratio is 10.5 or 12 to 1. This is more fuel than the engine actually needs: the chemically correct ratio is 14.6 to 1. But small engines, especially the hard-working two-strokes, use the surplus fuel as coolant.

Air, impelled by atmospheric pressure, passes through the filter and into the carburetor bore. About midway along the bore, the air stream encounters an obstruction, known as the *venturi* (Fig. 6-1). One might think that the obstruction would slow the air stream. Actually, quite the opposite happens— the air speeds up. To understand why, consider that as much air leaves the carburetor as enters it. Because the venturi is curved, it lengthens the path along which the air stream travels. Consequently, the air stream accelerates as its encounters the venturi. But nothing in nature is free, and the boost in velocity is purchased at the cost of pressure.

How fuel is induced to mix with the air stream depends upon the throttle angle (Fig. 6-2). At high speeds, the throttle is open or nearly so. Air moves

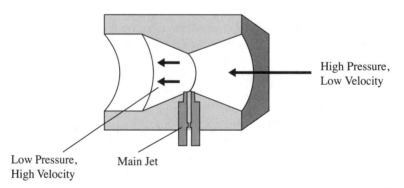

High Pressure,
Low Velocity

Low Pressure,   Main Jet
High Velocity

FIG. 6-1 *Air moving rapidly through a venturi loses pressure and gains velocity. The resulting low pressure "sucks" fuel through the main jet, which is mounted inside the main discharge tube.*

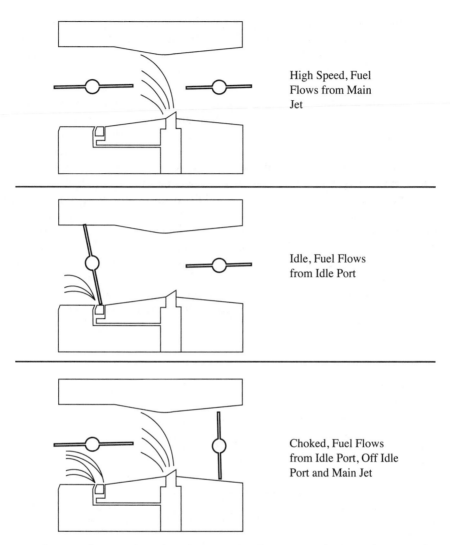

High Speed, Fuel
Flows from Main
Jet

Idle, Fuel Flows
from Idle Port

Choked, Fuel Flows
from Idle Port, Off Idle
Port and Main Jet

FIG. 6-2 *Carburetors have three basic modes of operation—high speed, idle, and cold engine. At high speeds, the throttle butterfly is open, permitting large flows through the venturi. Fuel enters the carburetor bore through the main discharge nozzle. At idle and low speeds, the nearly closed throttle functions as a primitive venturi to generate the vacuum that pulls fuel through the idle discharge port. As the throttle opens further, one or more transition ports come into play. A cold engine requires a rich mixture created either by a primer pump or as shown by a choke butterfly. When the choke is closed, the whole length of the carburetor bore comes under vacuum. Fuel flows from all discharge points.*

freely through the carburetor bore. Fuel, impelled by atmospheric pressure, passes through the main jet and discharges through a tube or nozzle situated at the venturi throat—the zone of least pressure and highest velocity. The pressure differential drives the fuel, and the velocity atomizes it into easily vaporized particles.

At idle, the throttle butterfly is almost closed. The tiny amount of air that passes through the bore cannot excite the venturi. But the butterfly itself acts as a venturi to draw fuel through the idle port. As the throttle opens wider, the butterfly uncovers a second, third, and sometimes a fourth port to smooth the transition between idle and midrange throttle. (On some carburetors, one off-idle port discharges air rather than fuel.) At between a third and half throttle, the venturi begins to develop vacuum, and the main jet adds to the fuel flow.

In a cold engine, much of the fuel condenses in the inlet tract. To compensate, cold-start mixtures are very rich. Some carburetors have a primer pump, but most use a choke butterfly upstream of the venturi. When engaged, the choke subjects the whole length of the carburetor bore to engine vacuum and all jets flow.

Fuel passages incorporate air bleeds that introduce bubbles into the fuel stream (Fig. 6-3). The resulting emulsion responds rapidly to changes in

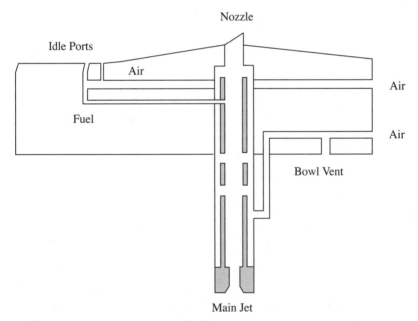

FIG. 6-3 *Air bleeds emulsify the fuel prior to discharge. The associated passages, including the cross-drilled holes in the nozzle, must be open if the carburetor is to function correctly.*

demand and resists siphoning. Diaphragm carburetors include a check valve in the main nozzle. As indicated earlier, the venturi is speed-sensitive, generating a vacuum only when the throttle is open. At idle and at low throttle angles, the venturi and the main nozzle come under atmospheric pressure. Were a check valve not present, incoming air would lower the level of waiting fuel in the nozzle (a condition that would result in stumbles on acceleration) and bleed into the idle circuit.

In order to hold mixture strength within narrow limits, a carburetor must have some means of regulating its internal fuel level. Most genset carburetors employ a float valve, similar to but far more sensitive than the valves used to control the water level in toilet tanks (Fig. 6-4). Fuel enters the carburetor through the needle and seat. As fuel is consumed, the float drops, permitting the needle to fall away from its seat. Fuel flows into the bowl. A fraction of a second later, the float rises and presses the needle closed.

Modern designs draw fuel from the center of the bowl so that the fuel supply will not be interrupted if the machine tilts a few degrees off horizontal. A central nut, often containing the main jet, secures the bowl to the carburetor casting. An O-ring seals the parts.

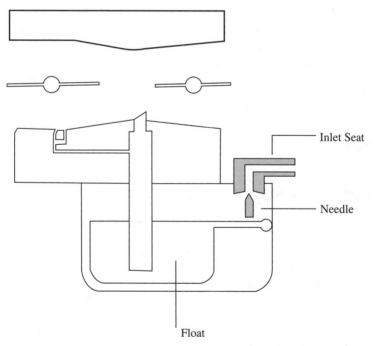

FIG. 6-4 *The float assembly, consisting of a hinged float and an inlet valve, maintains a constant fuel level in the carburetor.*

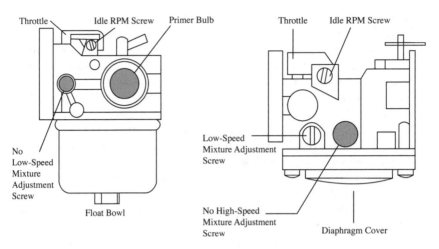

FIG. 6-5   *Float and diaphragm carburetors are instantly recognizable. The float carburetor on the left has an idle rpm adjustment. The diaphragm carburetor on the right permits adjustments to both idle rpm and idle mixture. Neither carburetor has an adjustable main jet.*

As one might expect, the delicate, watchlike float valve provides mechanics with plenty of work. Needles stick closed to deny fuel and, especially after being stored wet, stick open to flood the engine and dribble gasoline from the air filter. If that were not enough, the float bowl collects corrosion and varnish that finds its way into the main and idle circuits.

Diaphragm carburetors dispense with the float and instead use a flexible diaphragm to control the internal fuel level (Fig. 6-5). In other respects these carburetors work like float carburetors.

Tools and supplies

In the past, carburetors were soaked for hours in powerful metal cleaners. Plastic parts, many of them inaccessible, pretty much preclude the use of these cleaners. Most carburetor makers warn against their use. Tecumseh is a bit liberal in this regard and permits its carburetors to be immersed for 30 minutes, except for Series 11 and Series 11 Bridged carburetors (distinguished by black restrictor caps on the adjustment screws) that must not be dipped at all.

Shops now use aerosol cleaners, such as those marketed by Briggs & Stratton (PN Stratton 100041), Tecumseh PowerProducts (PN 696410), or any of the brands sold in auto parts stores. Exercise care with these products, which contain toluene and, often, methyl alcohol ketone (MEK). The Material Safety Data Sheets (MSDs) for most aerosol cleaners warn of blindness, birth defects, heart problems, and skin irritation. One company goes on record to say that death may occur from a single exposure to its carburetor cleaner. That's nice to know.

So be careful, work in a well-ventilated area, distant from ignition sources, and wear a face shield or goggles with side shields to protect your eyes from the spray. Gloves made of buna, polymer laminate, or nitrile are a necessity; toluene and MEK pass through the skin to target internal organs.

White vinegar works reasonably well and is much safer than commercial cleaners. Degrease with detergent and warm water, and immerse the carburetor in vinegar for 15 minutes or so. Rinse with water to remove all traces of vinegar, which is acetic enough to cause corrosion. I have used this technique on several Walbro float-type carburetors without obvious ill-effect, but the jury is still out on the more plastic-intensive diaphragm carburetors. A prolonged soak in vinegar will also remove rust from steel parts.

Serious carburetor work requires

- Hand tools. Walbro, Tecumseh, and Briggs & Stratton carburetors require inch-standard wrenches; other carburetors are metric. Several late-model Briggs carbs require Torx T-10, T-20, and T-30 wrenches, readily available at auto parts stores.
- Magnifying glass or jeweler's eye loupe.
- A powerful, narrowly focused light source.
- Compressed air. If you don't have a compressor, use canned air sold for cleaning computers.
- A soft-jawed vise.
- Spray carburetor cleaner and the items listed previously for personal protection when using these chemicals.
- A small ball-peen hammer.
- A hand-operated pressure pump, such as a Mighty Mite pressure/vacuum pump sold at auto parts stores or a Walbro PN 57-11 Pressure Tester. You can also adapt a radiator leak tester for carburetor work.
- Specialized carburetor tools. Tecumseh and Briggs & Stratton can supply the most popular tools, which are also available from Small Engine Suppliers (www.smallenginesuppliers.com). Walbro's PN 500-500 tool kit includes small chisels for removing Welch plugs, adjustment gauges for the company's products, and a slide hammer. To find a local distributor, contact the Walbro Engine Management, Aftermarket Division at 989-872-2131 (voice) or 989-872-7036 (fax).

Parts and materials

- Paper towels or lint-free rags.
- A way of removing oil-laden grime from external carburetor surfaces.
- Steel wool.
- Pipe cleaners.
- Fine-stranded copper wire for clearing jets and small ports. Copper does not upset the calibration by enlarging the orifices. Honda PN 07JPZ-001010B is the factory-tool equivalent.
- Nail polish (if it should be necessary to replace the Welch plugs).
- Carburetor kits come in two levels of completeness. The basic kit contains gaskets, O-rings, a needle and seat assembly, and diaphragms. The more comprehensive kits include nearly every part other than the carburetor casting.

With the exception of the inlet needle and seat, float-type carburetors do not wear out (few gensets accumulate enough hours to develop air leaks around worn throttle shafts). Failure is the result of varnish and corrosion. New parts are rarely needed. But it's a good idea to have an extra needle and seat, carb-mounting-flange and air-cleaner gaskets, and a couple of float-bowl ring gaskets on hand. Kohler bowl gaskets are among the most vulnerable.

Diaphragm carburetors are another matter. Replace the needle and seat, together with the soft parts—diaphragms and gaskets—whenever the carburetor is disassembled. Failure to do so is a recipe for frustration.

There appears to be immense profit in small-engine parts. One popular Mikuni needle-and-seat assembly goes for $30 and change. Parts are generally cheaper on the Internet and some Chinese clone manufacturers sell individual items. For example, DHGate.com offers a Yamaha EF2600 genset carburetor clone for $32.13 delivered to your home. The company accepts major credit cards.

Removal and installation

Remove the air cleaner, disconnect any wiring, and detach the springs, but leave the governor linkage in place for now. Make note of which holes the springs and links attach to. Some linkages are quite complex, and you may want to sketch the arrangement. Springs can be coaxed free with long-nosed pliers.

Remove the two mounting screws that secure the carburetor to the engine. As the carburetor is withdrawn, give it a twist to disengage the wire links.

Examine the gaskets and/or O-rings for damage and the telltale dark stains left by leaks. Run a finger into the carburetor bore. A sandy feel means that the air-cleaner element has failed or that the gasket between the air-cleaner assembly and the carburetor leaks.

Deposits on the bottom of the float bowl or under the metering diaphragm cover give an indication of the condition of the internal circuitry. Varnish, most of it anyway, can be removed, as can light rust on the float bowl. But water-induced corrosion of the sort shown in Fig. 6-6 means it's time to buy a new carburetor.

Running the mixture-adjustment screws down hard into their seats distorts both the seats and the tapered needles on float-type carburetors. The engine will run, but the distortion makes adjustment supersensitive. Overtightening adjustment screws on diaphragm carburetors has the same effect and can bring on a world of trouble by shearing off the tip, which remains lodged in the seat (Fig. 6-7). Finger-tight is plenty tight.

Installation is in the reverse order of disassembly, beginning with the throttle linkage. A small dab of sealant may be necessary to hold the gaskets in place, but use "stick-um" with discretion. You may have to take the carburetor off again, and it would be nice to do this without destroying the gaskets. And do not use silicon sealant, which dissolves in gasoline. Nor is it necessary to seal threaded fuel fittings with Teflon tape. Teflon encourages stripped threads and often ends up in the carburetor jets.

FIG. 6-6 *Carburetors in this condition are beyond repair.* Tony Shelby

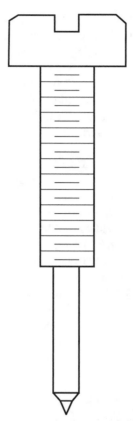

FIG. 6-7   *Diaphragm carburetor needle tips are vulnerable to twist-off. If this happens, you'll be shopping for a new carburetor.*

Cleaning

Clean the external surfaces with solvent, and disassemble the carburetor far enough to expose fuel and air passages. It is rarely, if ever, necessary to dismantle down to every last jot and tittle. Normally, one leaves the throttle and choke butterflies in place. Should the butterflies be removed, note any distinguishing marks as an assembly reference. Some butterflies have letters or code numbers on the outboard side; all have chamfered edges that seal against the carburetor bore. New (and practically unobtainable) screws should be used to secure the butterflies to their plastic shafts. Some parts, including many inlet fittings and Walbro fuel nozzles, are pressed into place and should not be disturbed. Nor should the expansion, or Welch, plugs be lifted other than as a last resort to clear internal passages.

In order to discourage tampering, modern carburetors either do not have an idle mixture-adjustment screw or limit range of the screw with a plastic restrictor cap. The caps can be drilled off or heated to soften the plastic and pried off. Spray carb cleaner into the fuel and air-bleed passages, and note where the spray exits to verify that no stoppages exist. You can make the same check with compressed air. Stubborn deposits in fuel lines can be removed with pipe cleaners, and jets cleared with strands of copper wire.

CAUTION: Wear protective gloves, work in a well-ventilated area, and keep the spray away from eyes and plastic tool handles.

Float carburetor service

Table 6-1 lists the most common malfunctions, most of which are caused by residue from evaporated fuel. Chapter 5 described checks to verify that the carburetor receives fuel.

Table 6-1
Typical float-type carburetor malfunctions

Symptom	Possible causes	Comments
Fuel leaking from the air filter	Inlet needle stuck open; float pivots binding	Paper filters swell into impermeability when wetted and must be replaced.
Overly rich mixture	Float set too high; improperly adjusted main jet	The correct spark plug for the engine should have a tan or light-brown firing tip. Excess fuel darkens the color. In extreme cases, rich mixtures blacken the exhaust with soot.
Overly lean mixture	Float set too low; improperly adjusted or clogged main jet; air leak downstream of carburetor at the flange or head gasket	Lean mixtures increase combustion temperature and lighten the color of the spark plug tip. In extreme, engine-threatening cases, the tip bleaches bone white, and the ground electrode shows blue temper marks. Two-stroke engines are far more susceptible to damage from lean mixtures than four-strokes.

(continued on next page)

Table 6-1
Typical float-type carburetor malfunctions (*continued*)

Symptom	Possible causes	Comments
Failure to start	Choke not fully engaged; primer inoperative; fuel stoppage—hung float, stuck inlet needle, clogged main jet	The choke must be fully closed for modern engines to start cold. Vintage automatic-choke butterflies are an exception and have 1/4- to 3/8-in. bore clearance when closed.
No or rough idle	Idle jet and/or idle feed circuit clogged; air leaks from a worn throttle shaft, carburetor flange gasket, or head gasket; choke not opening fully	
Stumble or flat spot on acceleration	Partial restriction in fuel supply to or at the main jet; float improperly set; air leaks downstream of carburetor (flange or head gasket)	
Rough, uneven running	Overly rich or lean mixture for reasons described earlier	Make certain that the problem is due to the carburetor and not the ignition system.
Engine runs for a minute or so and dies	Fuel stoppage; oil-level or oil-pressure sensor activated	Verify that the ignition continues to function as the engine coasts down to a stop.
Lack of power	Overly lean or rich mixture	

Needle and seat

Most inlet valves consist of a stainless-steel Viton-tipped needle that makes a fuel-proof seal against a brass fitting threaded into the carburetor body. A spring is sometimes used to pin the needle to the float or to dampen the seating impact (Fig. 6-8).

Flooding occurs if:

- A speck of sand or dirt lodges between the needle and seat. This fault often can be corrected by shutting off the fuel, removing the float bowl, and quickly opening and closing the fuel tap to clear the obstruction. Wipe up the spilled fuel before starting the engine.
- The needle sticks in the open position because of varnish on the needle or float pin. Clean as necessary.

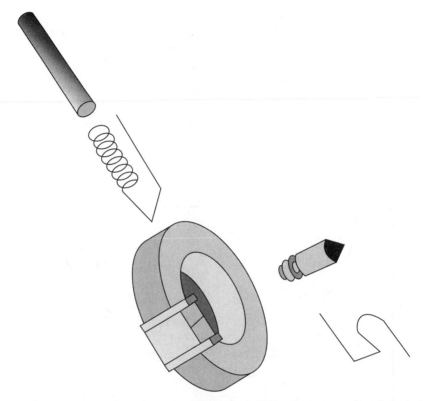

FIG. 6-8 *The needle may be secured with a hair spring to prevent it from hanging closed. The coil spring on the left absorbs seating shock.*

- The inlet fitting against which the needle seats has vibrated loose.
- The needle tip wears or distorts. Replace the entire assembly—seat ring gasket, seat, and inlet needle. Most seats are slotted for access with a properly fitting screwdriver.

A defective bowl gasket mimics flooding because the gasket must be sound to contain fuel slosh.

Should the inlet needle stick closed, no fuel will pass into the bowl. Clean the needle and float pivots.

Tecumseh 8 through 11 Series carburetors and most Briggs & Stratton units have elastomer inlet seats (Fig. 6-9). Remove the original with a piece of hooked wire, and press in the replacement part with a 5/32-in. punch and small hammer. The ribbed side of the seat goes down against the carburetor body. Some practice is required to install a malleable seat with enough force

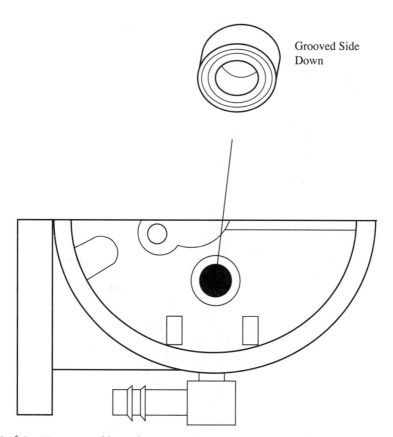

Grooved Side
Down

FIG. 6-9 *Viton assemble as shown.*

to seal and not so much that the seal distorts and leaks. If your luck is no better than mine, you'd need several replacement seats.

Some Mikuni carburetors have a pressed-in brass or steel seat, recognized by the absence of a provision for tool purchase. To extract these seats, thread an 8 × 32 machine screw into the fuel orifice. Prying on the screw releases the seat. Tap the replacement home with a flat punch.

Needle and seat assemblies can be tested by blowing on a short length of fuel hose connected to the inlet fitting. Air should pass through when the carburetor is held upright and should be blocked when the carburetor is inverted.

Float adjustments

Float height determines the level of fuel in the bowl, which, in turn, affects mixture strength (Fig. 6-10). Using long-nosed pliers and a small screwdriver

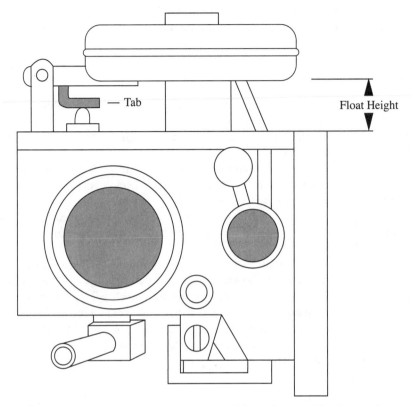

Tab

Float Height

FIG. 6-10 *Float height is most often measured from the toe of the float to the carburetor body.*

as a pry bar, bend the brass float tab as necessary. Do not raise the float by squeezing it down against the needle, an action that may distort the elastomer needle tip. Instructions that come with the rebuild kit describe the adjustment procedure. Most kits include a cardboard float gauge.

When all else has failed to provide sufficient fuel from an emissions-legal carburetor, mechanics have been known to raise the float height beyond specification. While not a magic fix, raising the level of fuel in the bowl contributes to a richer mixture.

A few carburetors have a second tab on the float or a plastic spacer to fix the drop limit, or how far the float falls when the bowl is empty. Excessive drop robs the float of mechanical advantage and can prevent the needle from seating. Mikuni float bowls often have a flattened side that must be assembled in the correct relationship with the float to permit the full range of drop. Tecumseh puts a ding in the bottom of the bowl for the same purpose.

NOTE: Walbro bowl nuts sometimes have two fiber seal washers, one in the normal position on the outer side of the float bowl and the other inside the bowl.

Jets

The main jet is located at either end of the discharge nozzle or is integral with the float hold-down nut (Fig. 6-11). If your carburetor has the latter arrangement, removing the nut and opening the clogged orifices often suffice to put the genset back into service.

Most discharge nozzles thread into the fuel pickup pedestal (the part of the body casting that extends down into the float bowl). Service these nozzles with a slender electrician's screwdriver. Clean the main jet and cross-drilled air bleeds (Fig. 6-12) with copper wire. Several carburetors, including Tecumseh's emissions compliant models, have nozzles that slip into place and seal with O-rings (Fig. 6-13).

Modern idle circuits draw fuel either from the main discharge nozzle or through a dedicated port on the lower end of the pedestal. A drilled passage in the pedestal conveys fuel up and into the idle discharge ports. Most carburetors incorporate an expansion plug that can be removed to give access to the fuel passage and/or the discharge ports. See "Diaphragm carburetor service" for further information. Verify that the idle circuit is open with compressed air or aerosol carb cleaner.

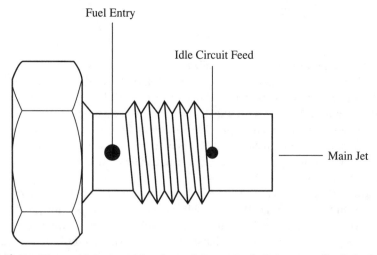

FIG. 6-11   *Tecumseh Series 12 bowl nuts have multiple functions, all of which are compromised by sediment that collects in the bowl.*

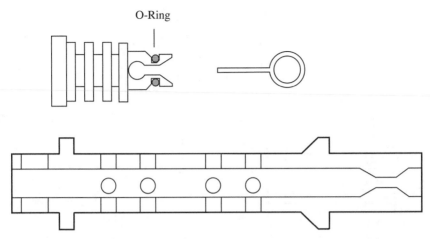

FIG. 6-12 *Nozzle and main jet for Honda GX120/140/160/200.270/340 and 390 engines.*

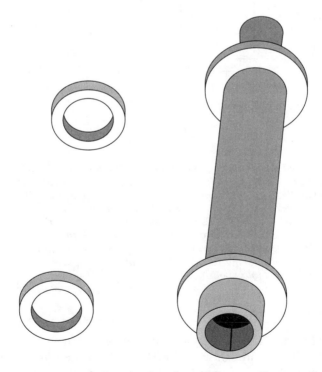

FIG. 6-13 *Tecumseh plastic nozzles, found on emissions carburetors (Series 640,000 and higher) mount with two O-rings. Be sure to extract the upper O-ring before replacing the nozzle.*

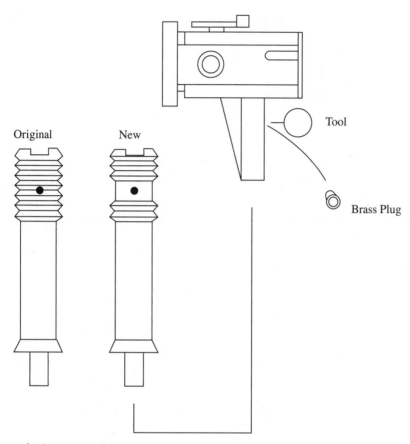

FIG. 6-14   *Walbro drills the idle feed port on LMB and LMG fuel nozzles after assembly. If you wish to use the original part and not the grooved replacement nozzle, remove the small brass plug located about three-quarters of the way up on the fuel pickup pedestal. Use a small tap with its tip ground flat to extract the plug. Thread the nozzle into the ¼ × 28 TPI (threads per inch) pedestal bore. Once the nozzle seats, run a small wire through the hole in the pedestal. Turn the nozzle back and forth. Assembly is correct when the wire probe passes through the cross-drilled hole in the nozzle and into the idle feed passage. Replace the brass plug, sealing it with a dab of nail polish.*

Walbro LMB and LMG carburetors, used on older Tecumseh engines, supply the idle circuit through a cross-drilled hole in the nozzle. Figure 6-14 shows how to index the nozzle with the idle passageway on assembly. Replacement nozzles have a groove that eliminates the need for alignment; however, when correctly aligned, the original part gives better idle quality.

Primer

The primer—a fuel pump that discharges into the carburetor bore—gives few problems. Should it fail, expect to find cracks in the elastomer bulb.

Diaphragm carburetor service

Diaphragm carburetors and fuel pumps receive their energy from crankcase pressure/vacuum pulses.

Crankcase pressure

Figure 6-15 shows the operation of a piston-controlled two-stroke engine of the type used on small, hand-carried gensets. As the piston rises, it evacuates the crankcase and opens the inlet port. In response to the partial vacuum, air and fuel enter the crankcase. A few milliseconds later, the piston reverses

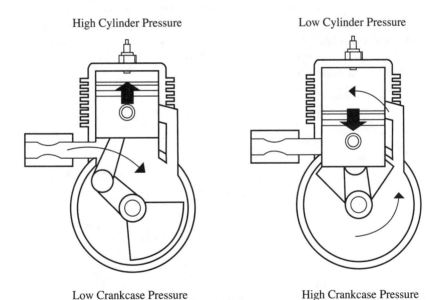

High Cylinder Pressure

Low Cylinder Pressure

Low Crankcase Pressure

High Crankcase Pressure

FIG. 6-15   *A two-stroke engine draws a mixture of air, fuel, and lubricating oil into the crankcase, which is then sealed and pressurized by the descending piston The engine illustrated uses the piston as a slide valve to control flow into and out of the crankcase. Other two-stroke engines employ an automatic reed valve for the same purpose. With either arrangement, the crankcase sees alternating pressure and vacuum pulses that can operate diaphragm carburetors and fuel pumps.*

course and retreats down the bore to close the inlet port. The falling piston pressurizes the now sealed crankcase. Further movement of the piston opens the transfer port to admit the fuel mixture to the upper cylinder, where it undergoes additional compression and ignition.

Diaphragm carburetor operation

The inboard side of the metering diaphragm is wetted with fuel and exposed to crankcase pulses (Fig. 6-16). The outboard side vents to the atmosphere. Pressure/vacuum pulses flex the diaphragm, unseating and reseating the inlet needle once each engine revolution, or 60 times a second at 3600 rpm. Fuel collected above the diaphragm passes through the main and idle circuits and discharges into the carburetor bore.

Most of these carburetors incorporate a pulse-driven fuel pump similar in operation to the metering diaphragm (Fig 6-17). A choke or, more commonly, a primer pump provides the extra fuel needed for cold starting.

As with any delicate device, successful repairs require a clean, well-lighted work space. Thoroughly clean the external surfaces of the carburetor before commencing disassembly. Some of the work, such as verifying that the inlet-needle spring is installed correctly in its boss or ensuring that adjustment needles have not been grooved by overtightening, can be assisted with a magnifying glass. Keep a notebook handy to record inlet needle and mixture screw settings and the relationship between the diaphragms and their gaskets.

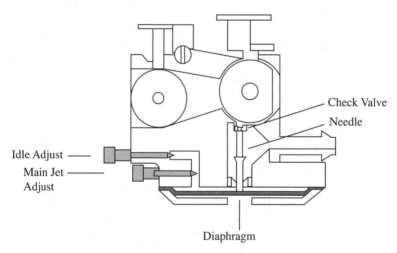

FIG. 6-16 *A diaphragm carburetor in its most basic form.*

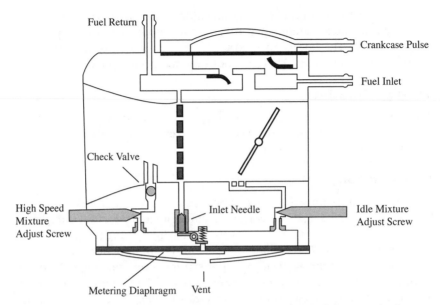

FIG. 6-17 *More sophisticated carburetors have a built-in fuel pump and a lever-actuated inlet needle.*

Plumbing

Should fuel or primer lines be confused, the engine will not start. If the primer mounts on the carburetor, two fuel lines are present. One line supplies the primer and carburetor, and the other returns surplus fuel to the tank.

When the primer mounts remotely, three lines are needed, one running between the carburetor and tank, another that returns fuel from the primer to the tank, and a third between the primer and the carburetor (Fig. 6-18). The latter is a suction line that draws fuel *through* the carburetor.

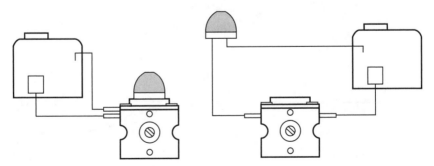

FIG. 6-18 *Primers integral with the carburetor draw from the tank and pump fuel to the carburetor. Remotely mounted primers draw fuel from the carburetor.*

If there's confusion about these connections, make up two fuel lines to the primer pump and submerge their free ends in a container of two-cycle fuel. Press the bulb. The line connected to the discharge side of the pump will reveal itself with bubbles. This line goes from the primer pump to the tank.

Troubleshooting

Table 6-2 lists some of the ways that diaphragm carburetor malfunctions affect engine performance. To make life interesting, most carburetor faults have multiple effects on performance.

Pressure test

Careful mechanics pressure-test a diaphragm carburetor before disassembly. Seal off the return line, if present, and connect a hand-operated pump, such as the Mighty Mite mentioned earlier, to the fuel-inlet fitting (Fig. 6-19). Apply 10 psi and no more. The carburetor should hold that pressure steady for at least 15 seconds. If the gauge needle drops prematurely, submerge the carburetor in solvent and retest. Bubbles coming out of the carburetor bore mean a leak at the inlet needle and seat. A trail of bubbles escaping from the diaphragm covers indicates a gasket or diaphragm leak. Carburetor castings also can leak, although such faults are rare.

Ultralight aircraft and kart enthusiasts are keen on modifying pop-off pressures. Most carburetors pop off, or open the inlet needle, at 10 to 12 psi. When the needle unseats, gauge pressure suddenly drops and then stabilizes at a

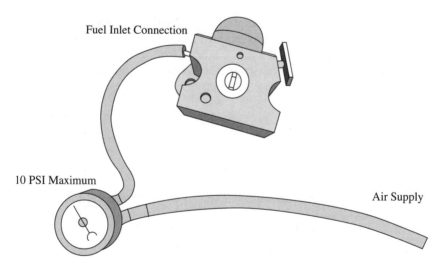

Fuel Inlet Connection

10 PSI Maximum

Air Supply

FIG. 6-19   *Pressure-testing saves time, frustration, and trips to the parts house.*

Table 6-2
Diaphragm carburetor malfunctions and engine performance

	Mixture	Hard or no start	Obvious flooding and/or fuel leaks	No or rough idle	Rich idle (exhaust smoke)	Fast idle (lean)	Flat spot or refusal to accelerate	Will not run at wide-open throttle	Low power at full throttle	Unstable rpm at full throttle
Clogged air filter	Rich	√		√			√	√		
Choke not closing or opening	Rich or lean	√	√		√ (if rich)		√ (if lean)	√	√	
Leaking carburetor flange gasket	Lean	√				√	√		√	
Idle circuit restricted	Lean	√		√		√	√			
Main jet restricted	Lean	√					√	√	√	√
Inlet needle stuck closed	Lean	Engine will not start								
Idle check-valve failure	Rich			√	√					√
Nozzle check-valve failure	Erratic	√		√					√	

(continued on next page)

Table 6-2
Diaphragm carburetor malfunctions and engine performance (*continued*)

	Mixture	Hard or no start	Obvious flooding and/or fuel leaks	No or rough idle	Rich idle (exhaust smoke)	Fast idle (lean)	Flat spot or refusal to accelerate	Will not run at wide-open throttle	Low power at full throttle	Unstable rpm at full throttle
Clogged main circuit air bleeds	Rich	√	√		√			√	√	
Stiff, cracked, or leaking metering diaphragm	Lean	√	√	√					√	√
Metering lever too high	Rich		√		√					
Metering lever too low	Lean					√		√		
Fuel pump leaks or otherwise malfunctioning	Lean						√		√	√

lower value to signal that the needle has reseated and is holding. A weaker needle return spring lowers the pop-off pressure for, it is said, some improvement in performance. However, carburetor manufacturers do not encourage experimentation in this area. All a genset owner needs to do is verify that the needle does in fact pop off at 10 psi, give or take a few pounds.

Mixture-adjustment screws

Emissions-compliant carburetors have limiter caps to restrict the range of mixture adjustment. Caps vary widely. Some can be snipped off and the needle unscrewed with long-nosed pliers; others can be grooved or heated and pried loose.

WARNING: The carburetor must be removed from the engine and dry before applying heat.

Once the caps are off, the problem remains of finding a tool to deal with the splined or otherwise oddly shaped screws. The EPA prohibits the sale of these tools to unlicensed technicians. Engine and genset manufacturers do not inventory them. But these tools do show up on eBay.

Some adjustment screws can be slotted with a Dremel tool and a cutoff disk; others respond better to a short piece of plastic fuel line, which is heated and slipped over the screw head. Or one can use a piece of rubber fuel hose with the appropriate ID. Mark the screws with a dot of nail polish, and count the number of turns to fully seated. Be gentle in order not to damage the needle points. High-speed and idle mixture screws may be of different lengths, but most have the same threads. To prevent confusion on assembly, place the screws in labeled plastic bags together with a notation of the number of turns the original setting was backed off from seated.

NOTE: The idle mixture screw is the screw closest to the engine.

Fuel pumps

Most carburetors have two diaphragms, with the one closest to the fuel inlet fitting acting as the fuel pump. Few problems are encountered with the pump, although the screen on the discharge side of Walbro and Zama pumps requires periodic cleaning. Use a pick to remove the screen, and install it with a wood dowel sized to fit. Note the relationship between the diaphragm and cover gasket. Some diaphragms go on first, followed by the gasket, and others reverse the order of assembly.

Metering

The metering diaphragm and inlet needle are the main focus of service work. Remove the screws that secure the metering chamber cover to the carburetor body. Make note of the position of the cover gasket relative to the diaphragm. Assemble these parts wrong, and the engine will not run.

Walbro, Zama, and their imitators actuate the inlet needle through a lever, which can be bent for adjustment. Before dismantling and losing the factory setting, measure the needle height (Fig. 6-20). Replacement parts may require adjustment. A Zama Z-gauge or the Walbro equivalent enables the adjustment to be made with precision. Too high a needle richens the mixture; too low a needle leans it (Fig. 6-21).

Idle circuits

Some Walbro carburetors cover the idle passages with a plate, secured by a screw. But the more common practice is to cover this and other drilled passages with pressed-in expansion, or Welch, plugs. Unless the carburetor is encrusted with gum and varnish, most mechanics leave sleeping dogs and factory Welch plugs lie. If the engine then shows evidence of fuel starvation, the carburetor must be disassembled for a more thorough cleaning.

Figure 6-22 illustrates one way of removing a Welch plug. Guide the chisel in at a shallow angle in order not to damage the carburetor casting. Another approach is to drill a small hole in the plug and extract the plug with a self-tapping screw and Vise-Grips. Clean the mating surface and tap the replace-

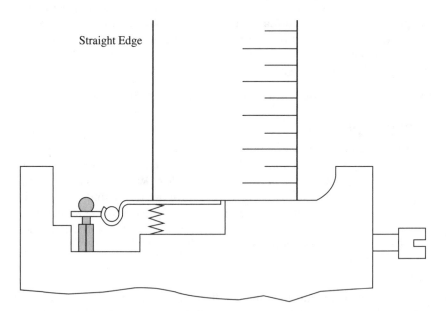

FIG. 6-20 *Lever height is measured with a straight-edge.*

Push Down

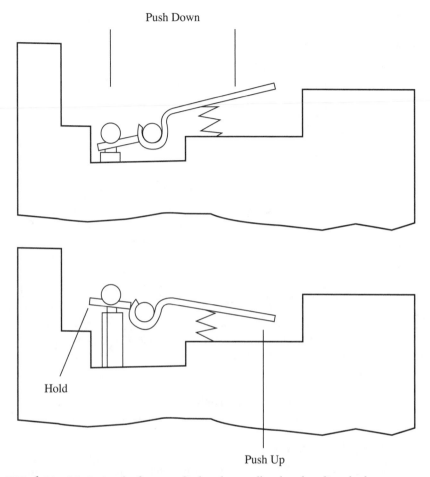

Hold

Push Up

FIG. 6-21 *Minimize the force applied to the needle when bending the lever.*

ment plug home, flattening the plug as its goes down. To improve security, seal the edges of the plug with nail polish.

Check valves

A ball or disk check valve blocks air entry into the nozzle. Air can enter during low-speed operation, when the venturi and the nozzle discharge orifice are exposed to atmospheric pressure. Some carburetors have a second disk-type or flapper check valve in the idle circuit. Main nozzle valves can be checked by alternately blowing and sucking on a piece of plastic tubing pressed against the nozzle inlet. Soft, flexible tubing with its end razor-cut

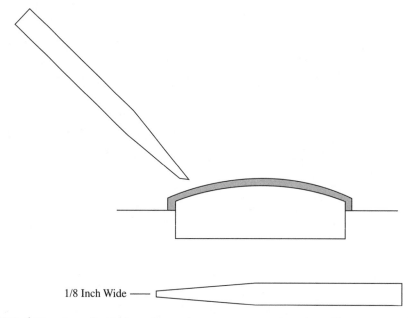

1/8 Inch Wide

FIG. 6-22 *A small chisel can be used to extract expansion plugs. The more comprehensive carburetor kits include replacement plugs, which are installed flat with a large punch and sealed with nail polish.*

dead square is required to make an airtight seal. The valve should open when pressurized from below and close with air drawn past it.

Final pressure check

It's a good idea to immerse the carburetor in solvent and apply 10 psi to the inlet fuel fitting as a final check. No bubbles should appear until pop-off pressure is reached.

Mixture adjustments: all carburetors

Pre-emissions carburetors have screws for both high-speed and idle mixtures. These regimes overlap—the idle circuit contributes a diminishing amount of fuel from idle to half throttle and beyond; the high-speed circuit begins to flow early and becomes the sole source of fuel at some point between half and three-quarters throttle. As a consequence, mixture adjustments are interrelated. Carburetors are usually adjusted by ear, but a tachometer gives more precise results.

Mark the screw heads with a dot of nail polish as a reference. Backing out the screws the same number of turns from seated, as originally found, will give the approximate setting. If the count has been lost, set the high-speed screw two or three turns out from seated and the idle screw two turns out. Run the engine under no load at top governed speed. Some mechanics adjust the idle mixture first; others, myself included, prefer to start with the high-speed adjustment.

1. Back off the high-speed adjustment screw in small, one-eighth-turn increments, pausing to give time for each adjustment to be felt. Continue until the engine falters at its rich limit. Note the position of the screw slot or paint mark.
2. Tighten the high-speed screw in the same manner, giving the engine time to respond to the progressively leaner mixtures. Stop when rpm drops at the lean limit. Note the position of the screw slot or the nail-polish mark.

CAUTION: Do not linger at the lean limit when working with two-stroke engines.

3. Back off the screw, splitting the difference between lean roll and the rich limit.
4. Repeat steps 1 through 3 for the idle adjustment.
5. Snap the throttle open with your finger. If the engine does not accelerate smoothly, back off the high-speed screw a small fraction of a turn and retest.
6. Check the idle adjustment. Sometimes we have to sacrifice idle quality for comfortably rich high-speed performance.
7. Verify that the genset runs at the rated rpm under full load. If not, tweak the adjustment screws as necessary.

Emissions-approved carburetors have fixed high-speed jets and a limited amount of idle-mixture adjustment, even with restrictor caps removed. One often ends by backing out the screws as much as possible.

Fuel injection

Electronic fuel injection (EFI) has been slow to penetrate the genset market and currently is available only on Kohler Aegis and Command Pro engines (Fig. 6-23). Advantages include better fuel economy, fewer exhaust emissions, and improved responsiveness.

ECH, Bosch, and Delphi have at different times supplied Kohler with its EFI systems. ECH builds injection systems for Toyota, Delphi for Harley-Davidson and other OEMs, and Bosch injection systems are nearly universal on European cars and trucks. Hardware and software differ, but all Kohler EFI systems operate primarily from voltage signals generated by throttle position

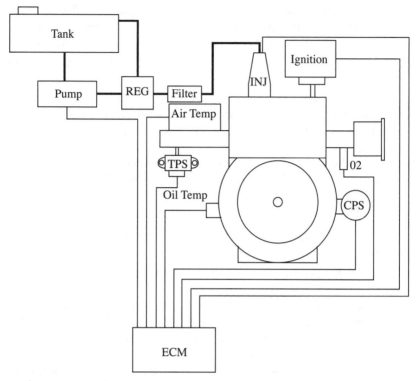

FIG. 6-23 *Kohler's fuel injection is a simplified version of automotive EFI systems. The throttle position sensor (TPS), crankshaft position sensor (CPS), oxygen sensor (O_2), and other components should be quite familiar to DIYers who work on their own cars. What's lacking is a catalytic converter and a mass airflow sensor.*

and rpm sensors. The electronic control module (ECM) integrates these signals to determine how long the injectors remain open and how much ignition advance to provide.

Other sensors enable the computer to increase injector pulse width during cold starts and in winter temperatures. The lambda, or O_2, sensor functions as a kind of reality check. This sensor continuously samples the oxygen content of the exhaust, signaling the ECM to make running changes to fuel delivery. Without an O_2 sensor, the ECM would be wholly dependent on its software, which cannot always anticipate real-world conditions. The computer also exhibits a degree of learning and stores its accumulated experience in volatile memory. Preserving this memory imposes a small drain on the battery when the machine is parked. If the battery goes dead or is disconnected, this memory is lost, and the engine will require a few minutes to recover its normal state of tune.

The factory provides free downloads of its service manuals on its website (www.kohlerengines.com). These manuals do a fairly comprehensive job of covering the various EFI systems, and the material need not be repeated here. But there are several points worth mentioning:

- Most malfunctions are due to pin and chassis ground connectors. Silicon dielectric grease helps keep corrosion at bay.
- Early EFIs had a grounding fault, since corrected with PN 24-452-01-S.
- Check that the fuel pump generates 37 to 40 psi with the appropriate Design Technology gauge. This tool can be purchased online at http://tinytach.com/tools.php. Depending on the supplier, the pressure regulator may be a stand-alone item or integrated with the fuel pump.
- Check for smoothly progressive changes in resistance as the throttle position sensor (TPS) opens and closes.
- Air leaks in the exhaust manifold upstream of the O_2 sensor result in excessively rich mixtures.
- A defective oil-temperature sensor incurs a fuel economy penalty because the computer will assume that the engine is cold and richen the mixture.
- A defective inlet air-temperature sensor results in hard starting, sudden shutdowns, and rough, uneven combustion.
- The crankshaft-position sensor (CPS) has been a perennial problem for automotive systems, and there's no reason to think that Kohler units should be more reliable.
- O_2 sensors degrade with age.

Diesel

Gensets powered by Lombardini (Kohler), Hatz, and Yanmar diesels make up a small but growing segment of the market. Most of these engines have a single cylinder and none more than two. All are air-cooled and have aluminum blocks and mechanical injectors. Figure 6-24 illustrates the basic fuel system.

To understand what diesel service work entails, some appreciation of the diesel cycle is needed. As air is compressed, it becomes hotter, as anyone who has used a hand pump to inflate a tire can attest. Diesel engines have compression ratios of about 20 to 1, which means that air ingested during the intake stroke is compressed to one-twentieth of its original volume. Cranking compression is on the order of 350 to 400 psi.

Fuel arrives late in the compression stroke when air pressure in the cylinder is already high. Genset engines develop injection pressures of between 2800 and 3800 psi compared to 40 psi for gasoline injection. To develop and

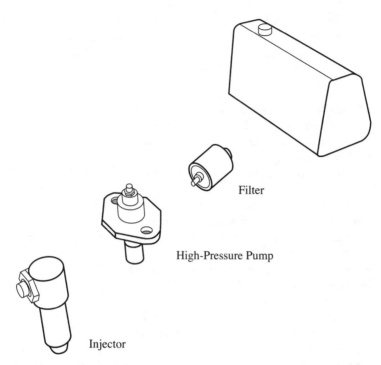

Filter

High-Pressure Pump

Injector

FIG. 6-24 *The basic diesel fuel system can be augmented with additional stages of filtration and a transfer pump to move fuel from the tank to the high-pressure pump.*

contain these pressures, fuel pumps and injectors exhibit extremely close fits with lapped, rather than machined, sealing surfaces. For these components to live, the fuel must be exceptionally clean and free, in so far as possible, of water. Some manufacturers warn against refueling in the rain. Fuel filters, which block particles as small as 7 micron in diameter [(1micron = 0.001 mm), clog rather quickly, and fuel tanks must be replaced at the first sign of rust.

Injection commences at about 20 crankshaft degrees before top dead center, but ignition lags behind injection. Some appreciable time is required for the newly injected fuel to collect heat and vaporize. When ignition does occur, the collected fuel goes off almost instantaneously with correspondingly rapid rises in cylinder pressure. These high pressures account for the "diesel knock" one hears at idle.

A diesel engine needs fuel, air, and compression to run. Air presents few difficulties, and compression tends to remain stable over the duration of engine life. Fuel is the wild card, the source of nearly all malfunctions (Table 6-3).

Table 6-3
Diesel troubleshooting

Symptom	Probable causes
Black or gray exhaust smoke under load.	*At high speed:* • Clogged air filter • Dirty injector, especially if power is down *At medium and high speed:* • Injector timing retarded *At low and medium speed:* • Injector timing advanced (engine may be noisier than normal) • Clogged/restricted fuel lines *At all speeds, but most apparent at low and medium speeds, engine may be difficult to start:* • Weak cylinder compression *Puffs of black smoke at all speeds, sometimes with a blue or white component:* • Sticking injector
Whitish or blue smoke at high speed and under light load, especially when engine is cold. As temperature rises, the exhaust smoke darkens. Power loss across the entire rpm band.	Injector pump retarded
Whitish or blue smoke under light loads after the engine reaches operating temperature. Knocking may be present.	Leaking injector
Blue smoke under acceleration, on startup, or after a long period at idle. Smoke may disappear when running at steady speed.	Worn valve oil seals and/or valve guides
Blue smoke at all speeds, loads, and operating temperatures.	Worn piston rings/cylinder bore
Increases in crankcase oil level.	Fuel pump leaking into crankcase.
Slow cranking speed.	Generator bearing failure Engine bearing failure Electric start: • Low state of battery charge • Corroded or loose battery cables • Bad cable ground • Starter motor malfunction

(continued on next page)

Table 6-3
Diesel troubleshooting (*continued*)

Symptom	Probable causes
No or hard start.	Insufficient fuel: • Fuel tank empty • Manual or solenoid fuel valve closed • Air in fuel system • Clogged fuel filter • Injector failure • Low pump pressure and/or delivery Contaminated fuel Clogged air filter Weak cylinder compression: • Insufficient cranking rpm • Blown head gasket • Worn piston rings/cylinder bore • Leaking valves • Air leak at injector mount
Engine does not achieve full rpm at no load.	Governor out of adjustment
Surge, often most pronounced at idle. Engine may shut down at random.	Air in fuel system
Engine starts and runs normally but shuts down within minutes.	Air in fuel system Clogged fuel filter Clogged air filter Fuel return line restricted
Engine starts normally but misfires.	Air in fuel system Clogged air filter Clogged fuel filter Malfunctioning injector Weak cylinder compression
Low power.	Air in fuel system Contaminated fuel Insufficient fuel supply: • Clogged fuel filter • Injector malfunction • Low pump pressure • Fuel leak Clogged air filter Heavier than normal generator loads Governor out of adjustment

Wet stacking

Some owners buy an oversized genset in the belief that surplus power ensures reliability. This is a defensible, although expensive, strategy for spark-ignition generators. But diesels are workhorses, happiest when running at 60 to 75 percent of rated load. If the engine spends much time at idle or under light loads, cylinder temperatures drop and combustion suffers. Cylinder bores glaze and score, piston rings stick, and injector tips carbonize. Water and other by-products of combustion collect in the crankcase to the detriment of the bearings and other friction surfaces. Raw fuel collects in the exhaust pipe or stack. If the carbon deposits on the exhaust outlet leave a wet smudge on your finger, the engine is in trouble.

Running at rated load for several hours burns off much of the carbon and may free the rings. In many cases, the engine can be returned to service. But the effects of chronic, long-term wet stacking cannot be so easily undone. A comprehensive overhaul is the only recourse.

Safety

Contact with diesel fuel is unavoidable—mechanics use the fuel as a solvent, prime pumps with it, and inevitably get it on their clothes. DIYers probably need not worry because exposure is limited. But for the record, diesel fuel passes easily through the skin with biological impacts that have yet to be determined more than a century after the prototype engine ran. One less-than-definitive study detected a correlation between prostate cancer and fuel exposure. Nitrile or Viton gloves appear to give the best protection; neoprene is problematic, and vinyl or butyl rubber soaks up diesel like blotters.

Injector spray is another matter. Spray or leak streams from the high-pressure side of the injector pump easily penetrate the skin to cause blood poisoning or, at the very least, leave a "diesel tattoo." Strikes to the eyes don't bear thinking about. Wear goggles or a face shield, and when testing injectors, dampen the spray with a rag or paper towels.

Tools

Routine maintenance requires no more than a set of metric wrenches. Readers who want to delve more deeply into fuel systems will need to fabricate tools. Little by way of small-engine tooling is available, and what tools there are tend to be quite expensive. For example, a Yanmar injection holder—a steel plate with slots cut in it—retails for $130.

Fuel system checks

Restrictions upstream of the high-pressure pump reduce the volume of fuel delivered, and the presence of air makes delivery sporadic, often to the extent that the engine refuses to start. Air enters whenever fuel retreats, such as when a filter is changed or the tank runs dry. If the engine suddenly runs rough, shuts down, or becomes difficult to start, suspect air in the fuel system. Another indication of air intrusion is froth in the return line that runs between the injector(s) and tank.

If the engine fails to start:

1. Verify that there is clean, uncontaminated diesel automotive fuel in the tank.
   - Shut off the manual valve at the tank.
   - Disconnect the flexible hose that supplies the high-pressure pump. Once the clamp is undone, twist the hose to break the connection. A heat gun can assist in loosening old hoses.
   - Open the valve. If there is no stoppage, fuel should flow out the open line.
   - Reconnect the feed line.
2. Loosen the high-pressure line at the injector a half turn or so (Fig. 6-25). Briefly crank the engine to verify that the injector receives fuel and to bleed any air present between the pump outlet and injector.

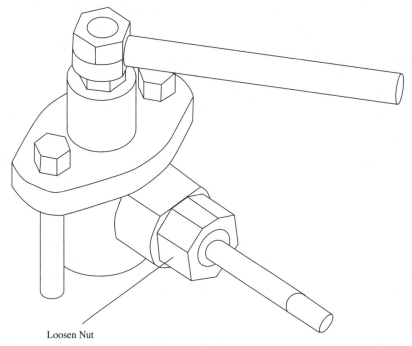

Loosen Nut

FIG. 6-25 *Loosen the high-pressure union at the injector to check for fuel delivery and to purge air from the system.*

3. If the pump receives fuel and the injector does not, repair or replace the pump.

This approach usually works. One can sometimes speed things up by priming the pump with a squeeze bottle of fuel or hand pump.

When air intrusion is chronic, run the edge of a business card around the fuel line connections. If the card discolors with absorbed fuel, a leak is almost surely present. Tighten the connection; should the leak persist, replace the copper sealing washer. These washers are not always easy to find and you may have to make do with the original part. Fine-grit emery paper supported by a piece of plate glass or a machine worktable will take out most of the scores. As further insurance against leaks, anneal the washer by heating it with a propane torch and letting it cool slowly in air.

Injectors

Small-engine injectors are spring-loaded, normally closed valves that open at nozzle opening pressure (NOP). Fuel enters the injector through a hard line, discharges through multiple orifices in the nozzle tip, and the surplus returns to the tank through a flexible hose.

Injector body   The body includes a flange or other mounting provision and connections for the high-pressure input and return lines. An O-ring provides a gas seal between the injector body and the cylinder head.

Internal parts include drilled passages for fuel ingress and return, and a heavy coil spring, together with a shim stack or other means of adjusting spring tension and the resulting NOP.

Nozzle   The nozzle threads into the injector body and opens to the combustion chamber (Fig. 6-26). Yanmar-style nozzles use spacer washers to define the depth of nozzle extension. If you take the injector apart, be sure to replace all the spacers on reassembly. Performance suffers if a spacer is left out.

Fuel radiates outward from four or more orifices in the nozzle tip in a nearly horizontal spray pattern. The tiny droplets vaporize and burn before making contact with the cylinder walls. Lister engines and "Listeroid" clones inject into an antechamber adjacent to the combustion chamber. These dated, but very practical, engines use single-hole nozzles to deliver a narrow stream of relatively large fuel droplets deep into the antechamber.

Needle   Fuel pressure acting on the needle shoulder (the upper "lift surface" in Fig. 6-26) raises the needle off its seat. Injection, accompanied by a pressure drop, commences. The additional surface provided by the needle flank (the lower lift surface) holds the needle open under the reduced pres-

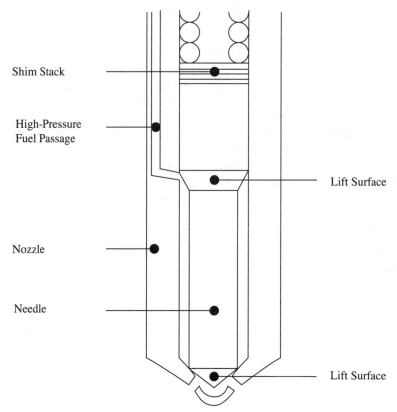

FIG. 6-26 *Fuel pressure acting on the lift surfaces opens the injector against spring tension.*

sure. As pressure continues to drop, the balance of forces shifts toward the spring, and the needle seats.

Injector testing can be done on the engine. Disconnect the high-pressure line, nudge it out of the way being careful not to bend it, and undo the injector hold-down bolts. Extracting the injector on a high-hour, weather-beaten engine can present a challenge. Flood the sealing surface with penetrating oil, and twist the injector back and forth. Needless to say, the O-ring must be renewed on reassembly.

Once the injector is retrieved, place a rag under the injector tip, reconnect the high-pressure line, and crank.

WARNING: Keep hands and face clear of the high-pressure spray.

Figure 6-27 illustrates the desired spray pattern for multihole nozzles. A diesel repair facility can make more comprehensive tests, but the money

Normal   Distorted

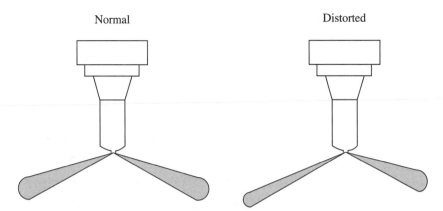

FIG. 6-27 *The spray pattern should be symmetrical and defined, extending for essentially the same distance from each orifice.*

spent on testing might be better applied to the purchase of a new injector. As with spark plugs, it's useful to have one or two spares on hand.

Polish carbon deposits from the nozzle tip with a softwood block, and use fine piano wire to clean the orifices. If you opt to disassemble the injector, inspect the needle and nozzle seat with a magnifying glass. A fuel-wetted needle should rotate easily without binds and fall of its own weight when the nozzle is held off the vertical.

High-pressure pump

Figure 6-28 shows a single-plunger Yanmar-style pump of the sort most often used on portable generators. An eccentric on the engine camshaft drives the plunger through a tappet. Fuel provides lubrication for the plunger and engine oil for the drive mechanism. Pumps for multi-cylinder engines are far more complex. Bosch-pattern pumps have a dedicated plunger for each engine cylinder, a camshaft and a self-contained lubrication system.

A check valve, known as a delivery valve, opens to permit flow to the high-pressure line and closes abruptly when pump pressure falls. The sharp cutoff of injector pressure helps to prevent nozzle after-dribble. Although failure is rare, the valve can be tested with a vacuum pump.

Some pumps are fitted with a normally closed solenoid-operated valve that enables the engine to be shut down by remote control. The valve should click when energized. If it fails to do so, either replace the solenoid or disable the valve by removing the solenoid plunger.

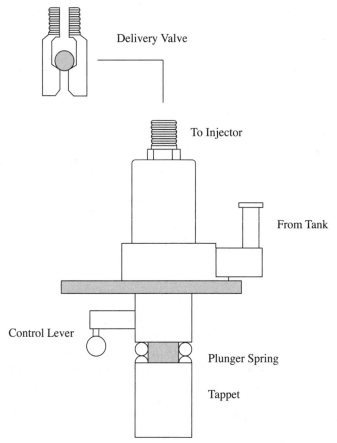

FIG. 6-28   *Most small injection pumps follow the Yanmar example.*

Timing

The height of the pump above the camshaft eccentric determines injection timing. The higher the pump, the later the onset of injection. Lombardini uses washer-like shims between the roller tappet and pump body; Yanmar and most other pumps use brass shims under the pump flange gasket. Gasket paper can substitute for the latter. Small, fractional-millimeter differences in shim thickness have a profound effect on timing and engine performance.

A decrease in the height of the shim stack is appropriate when camshaft wear delays timing or when operating at high altitude. Unless specially ordered, gensets come with the timing set for sea-level operation. Elevations of 5000 ft and greater call for a small advance to compensate for the increased ignition lag caused by reduced atmospheric pressure. Advancing the timing

restores some of the performance lost at altitude, but the engine will be nois-
ier and may emit visible exhaust smoke.

The control lever terminates in a ball that fits into a fork on the rack. These
parts must mate when the pump is mounted. An inspection slot on Yanmar-
style pump flanges assists in alignment; align other pumps by advancing the
rack and control lever to the full-speed position prior to installation.
Installation goes easier if the flywheel is rotated to put the pump plunger low
on the camshaft eccentric.

If you are at sea about timing—if shims have been lost or a clone pump of
unknown dimensional accuracy has been fitted—time to the flywheel marks.
The flywheel will have at least two marks on its rim. In the normal direction
of rotation, the first mark to come into alignment with the pointer indicates
the onset of injection. This is the timing mark. The second mark means that
number one piston is at top dead center.

To time to the flywheel mark:

1. Remove the shroud so that timing marks are visible.
2. Replace the high-pressure fuel line at the pump with a fabricated gauge
   such as illustrated in Fig. 6-29.

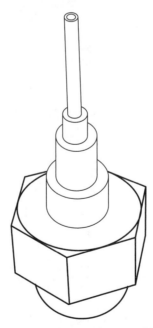

FIG. 6-29 *An extremely sensitive gauge can be assembled from a fuel-connection
nut and various sizes of plastic tubing used as reducers. A transparent tube from
an aerosol spray can acts as a sight glass. Seal the tubes with silicon.*

3. Affix a large wrench on the flywheel nut, and bar the engine over to bring the fuel level to the midpoint of the gauge.
4. Continue to rotate the flywheel, slowing as the timing mark approaches the pointer.
5. Keep a sharp eye on the fuel level. It should begin to rise at the moment the timing mark indexes with the pointer. If you overshoot the mark, back off the flywheel 20 degrees or so, and try again, this time working more slowly.
6. Repeat the test several times to verify.
7. Shim the pump as necessary. As one might suppose, the effect of shim thickness on timing varies with engine make and model. But the relationship is dramatic. For example, each 0.1-mm (0.004-in.) change in the thickness of Lombardini shims changes the timing by 1 crankshaft degree.

A more precise method of determining the onset of injection is to remove the delivery valve and measure plunger movement with a dial indicator (Fig. 6-30).

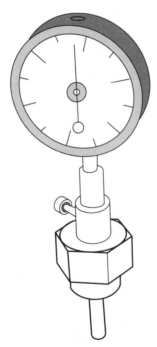

FIG. 6-30 *A dial indicator provides the most reliable information on plunger movement. The arrangement shown mounts the indicator in a sleeve soldered on a fuel-line nut. For this set up to be practical, threads for the fuel-line nut must be present at the pump outlet when the delivery valve is removed.*

7

Major engine repairs

According to the U.S. Consumer Product Safety Commission (CPSC), the light-duty yard and garden engines that power most gensets have an operating life of 500 hours, or twenty-one 24-hour days. Name-brand diesel and quality spark-ignition engines do much better, but no portable generator can provide long-term security. For that, one needs a standby generator, preferably one powered by a multicylinder, liquid-cooled diesel.

What I've tried to do in this chapter is to explain engine repairs in ways that factory shop manuals, which are written for professionals, do not. Readers who prefer to farm the work out still need to understand what is entailed because the customer is the quality-control inspector. Mechanics and machinists make mistakes, take shortcuts, and make tradeoffs. I don't know of a single engine rebuild that could not have been done better.

Some basics

Readers new to the subject need first of all to get a mental picture of how engines work. An internal combustion engine must induct air and fuel, compress the mixture, ignite it and exhaust the byproducts of combustion. Four-stroke or four-cycle (the terms are used interchangeably) engines perform these actions in four strokes of the piston or two full crankshaft revolutions. Two-stroke engines telescope events into a single crankshaft revolution. Figure 7-1 illustrates the four-cycle principle as used by spark-ignition engines, Fig. 7-2 does the same for diesel engines, and Fig. 7-3 shows two-cycle operation.

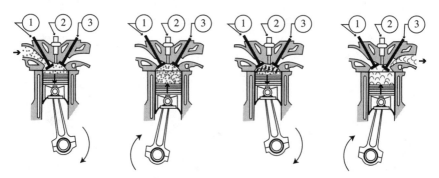

FIG. 7-1 *Four-cycle spark-ignition engine operation. (Left) The falling piston draws air and fuel through the open intake valve (1). (Center, left) Both the intake (1) and the exhaust (3) valves are closed, and the piston begins to rise on the compression stroke. (Center, right) The moment of ignition as the spark plug (2) fires to initiate the expansion, or power, stroke. (Right) The piston rises on the exhaust stroke. Thus four strokes of the piston, accomplished by two full revolutions of the crankshaft, are required to complete the cycle.*

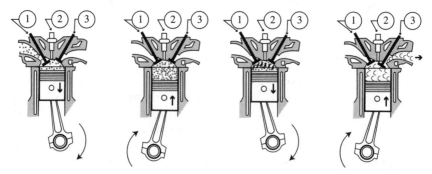

FIG. 7-2 *Four-cycle diesel engines employ the same intake, compression, expansion, and exhaust strokes shown in Fig. 7-1. Diesels, however, take in only air during the intake stroke. Fuel enters under high pressure through an injector (2) late in the compression stroke. The heat of compression then ignites the air-fuel mixture.*

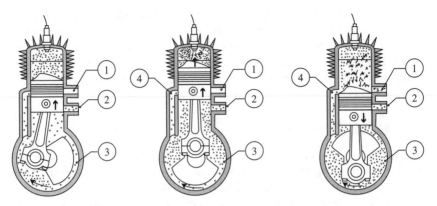

FIG. 7-3 *Two-cycle operation involves simultaneous events above and below the piston. (Left) The piston rises on the compression stroke and depressurizes the crankcase. (Center) Ignition. At this point in the cycle, the piston has uncovered the inlet port (2), which admits a fresh air-fuel charge into the crankcase (3). (Right) The piston descends on the expansion stroke, an action that both pressurizes the crankcase and opens the transfer port (4). Air and fuel pass through this port to fill the cylinder area above the piston. The rising piston then completes the cycle by compressing the mixture in the cylinder and de-pressurizing the crankcase.*

Because the crankcase is part of the induction tract, two-stroke engines receive lubrication by way of oil mixed with the fuel. The three-port engine illustrated is one of several varieties. Other two-cycle engines control crankcase filling with a reed or disk valve.

What's needed

While some frequently used specifications are included here, a factory shop manual is an essential tool. Many of these manuals can be downloaded at little or no cost from websites listed at the back of this book.

In addition to usual assortment of hand tools, you will need a few specialized tools, such as a torque wrench and measuring instruments. These tools are described subsequently in the contexts they are used.

Imported engines require metric-standard tools; those manufactured in this country conform (with rare exceptions) to the inch standard.

It's axiomatic that dirt on external engine surfaces migrates by way of hands and tools to internal surfaces when the engine is opened. Some DIY mechanics begin by spraying down the engine in a car wash; others rely on Gunk—a product of less-than-stellar environmental impact that turns oil and grease into a kind of slimy soap that rinses off with water. Wet down the engine with the cooling shroud in place and the paper air filter element removed. Major castings can be hosed down again after disassembly.

Internal parts respond to kerosene or Varsol. Gasoline is far too hazardous to use as a solvent. Readers might want to follow the example of professional mechanics and wear nitrile gloves for protection against gasoline, diesel fuel, and petroleum-based solvents, all of which pass readily through the skin. Ziplock plastic bags keep fasteners and other small parts organized.

Torque limits

Snap-On has found that mechanics overtighten bolts and nuts by 20 to 30 percent. Overtightening elongates the threads and can result in strip-outs, especially in aluminum. Undertightening defeats the purpose of the fastener, which is to create a clamping force between two or more parts.

Table 7-1 lists torque limits for common metric fasteners. The table assumes that threads are assembled dry, without lubrication. As a bolt is tightened, about 90 percent of the twisting force reported by the torque wrench represents friction. The 10 percent that remains is the clamping load. Lubricating a fastener can easily reduce friction by 5 percent. That tiny reduction almost doubles the clamping force. Unless the manufacturer indicates otherwise, do not use lubrication on threaded joints. By the same token, dirty, corroded, or stretched threads increase friction and result in undertightening.

Head bolts and other critical fasteners are sometimes torqued to a limit and then rotated a specific number of degrees tighter. The angle of rotation gives a better indication of clamping force than torque, which is so compromised by friction.

NOTE: Apply 80 percent of the listed torque value when the fastener makes up to aluminum.

Because most gensets originate overseas, torque specifications are nearly always given in newton-meters (N-m). Foot-pounds (ft-lb) or inch-pounds (in.-lb) have become increasingly provincial.

1 ft-lb = 12 in.-lb
1 ft-lb = 1.36 N-m
1 in.-lb = 0.8 ft-lb or 0.11 N-m
1 N-m = 0.74 ft-lb or 8.88 in.-lb

Torque limits for critical components—connecting rods, cylinder heads, flywheels, and crankcases—are make- and model-specific and differ from the general specs listed in Table 7-1. Specifications can be obtained from the manual or from the Outdoor Power Information website (http://outdoor powerinfo.com/).

Table 7-1
General-purpose torque limits, metric hex and
recessed-head (Allen) fasteners

Shank diameter (mm)	Head marking [ft-lb or, when indicated, in.-lb (N-m)]			Hex wrench size	
	8.8	10.9	12.9	Inch	Metric
M3	11 in.-lb (1.2)	14 in.-lb (1.6)	19 in.-lb (2.1)	7/32	5.5
M4	26 in.-lb (2.9)	36 in.-lb (4.1)	43 in.-lb (4.9)	9/32	7
M5	53 in.-lb (6.0)	6 (8.5)	7 (10)	5/16	8
M6	7 (10)	10 (14)	13 (17)	—	10
M8	18 (25)	26 (35)	30 (41)	1/2	13
M10	35 (48)	51 (69)	61 (83)	11/16	17
M12	62 (86)	88 (120)	107 (145)	3/4	19
M14	99 (135)	140 (190)	169 (230)	7/8	22
M16	157 (210)	217 (295)	262 (355)	15/16	24

Weekend mechanics can get by with an inexpensive 3/8-in. drive, beam-type torque wrench. Ironically, these simple tools that measure torque by the deflection of a beam work longer and more reliably than the click-type or dial-indicating wrenches favored by professionals. Tighten fasteners smoothly with your hand centered on the handle.

Evaluation

It is always useful to make some preliminary evaluation of the problem. Table 7-2 lists the mechanical faults that piston engines are susceptible to. The table assumes clean fuel and for spak-ignition engines a functional ignition system and carburetor.

Excessive compression

Overhead-valve (OHV) engines with automatic compression releases often develop too much compression to start. If this happens, the flywheel locks as the piston approaches top dead center (TDC) on the compression stroke. Rewind starters bind and electric starters burn out, often destroying the starter relay in the process. If removing the spark plug restores starter function, excessive compression is almost surely the problem.

Table 7-2
Mechanical malfunctions

Engine Type	Symptom	Probable causes
All engines	Crankshaft locked	• Jammed starter mechanism • Rust-bound rings • Hydraulic lock—oil or fuel accumulations above the piston • Parted and jammed connecting rod • Severely bent crankshaft
	Crankshaft drags	• Bearing damage caused by lubrication or cooling-system failure • Bent crankshaft
	No or weak cylinder compression	• Blown head gasket • Worn, stuck, or broken rings • Worn or scored cylinder bore • Holed piston • Broken connecting rod • Broken crankshaft
	Excessive vibration	• Bad motor mounts • Bent crankshaft
	Loss of power	• Weak compression • Clogged spark arrestor • Clogged exhaust ports and/or muffler (two-stroke engine) • Leaking crankshaft seals (two-stroke engine) • Leaking carburetor flange gasket
Four-cycle engines	Erratic idle	• Stuck crankcase breather valve • Leaking valves
	Misfiring, stumble under load	• Improper valve clearance • Weak valve springs
	Crankcase breather passes oil	• Leaking gasket • Clogged breather • Clogged drain in breather box • Excessive blow-by due to worn rings, piston, and/or cylinder bore
	High oil consumption	• Worn piston rings and/or piston-ring grooves • Stuck piston rings • Worn piston and cylinder bore • Clogged piston oil-drain holes • Worn valve guides • Faulty crankcase breather • Oil leaks

Table 7-2
Mechanical malfunctions (*continued*)

Engine Type	Symptom	Probable causes
	Power diminishes until engine shuts down a few minutes after startup	• Insufficient valve lash • Valves fail to seat as engine warms and parts expand
	Excessive compression that makes it impossible to turn the engine over with the rewind starter; engine spins freely with spark plug removed	• Excessive valve lash (when an automatic compression release is fitted) • Failure of the compression release
Two-cycle engines	No or little crankcase compression	• Leaking main-bearing seals • Failed reed valve (when present)
	Misfiring	• Leaking carburetor flange gasket • Leaking main-bearing seals

Automatic compression releases unseat the intake or exhaust valve during cranking to bleed off the excessive pressure (Fig. 7-4). While these devices can fail, the usual problem is excessive valve clearance. Valve clearances are sometimes included in the owner's manual and, of course, will always be found in the shop manual.

NOTE: In an emergency the engine can be started by doubling up on the head gaskets. The loss of compression has a small, but perceptible effect on full-throttle power..

NOTE: Kohler Aegis and some Command Pro gasoline engines and all modern Lombardini diesels have hydraulic valve lifters that automatically maintain the correct lash.

With few exceptions, side-valve engines have no provision for adjustment. One must grind the tip of the valve stem to increase lash or resurface the valve seat to reduce lash. Most DIYers farm out the adjustment to an automotive machinist.

FIG. 7-4 *Compression releases take many forms, probably because of patent restrictions. Briggs & Stratton has been most ingenious. Some of the company's engines have a ramp on the cam that interrupts intake-valve closing to leak compression during cranking. At running speeds, the compression leak becomes tolerable. The more common approach is to use a weighted or spring-loaded lever that holds the valve open during cranking and closes under centrifugal force when the engine starts. The lever tip wears quickly and, when a spring is involved, it too is vulnerable. Attempts to reface flippers with a weld bead have had mixed results.*

To adjust OHV lash:

1. Place a rag under the valve cover to catch the oil spill. Remove the cover.
2. Turn the engine to top dead center on the compression stroke. A wooden pencil inserted into the spark plug boss can be used to track piston motion. Hold the pencil at a shallow angle so that it does not become trapped between the piston and combustion chamber roof as you slowly turn the flywheel by hand.
3. At TDC (top dead center, or the upper limit of piston travel), both valves will be closed. Late-model Briggs & Stratton engines are an exception. Because of the automatic compression release, the piston must be positioned 1/4 in. past TDC. Make the adjustment with a flat-bladed feeler gauge as illustrated in Figs. 7-5 and 7-6. The work goes easier if the adjustment is bracketed. For example, a 0.004-in. blade will easily slip through a gap of 0.005-in. A 0.006- or 0.007-in. blade will exhibit perceptible drag.
4. Rotate the flywheel two full revolutions, and check the adjustment.

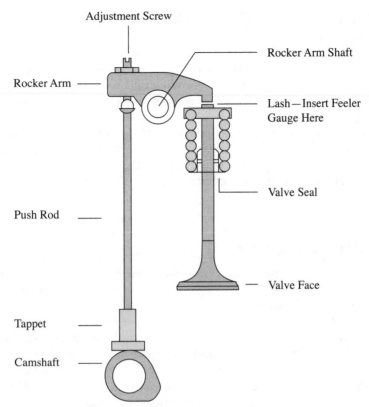

FIG. 7-5 *Lash for shaft-mounted rocker arms is measured at the interface between the rocker and valve-stem tip.*

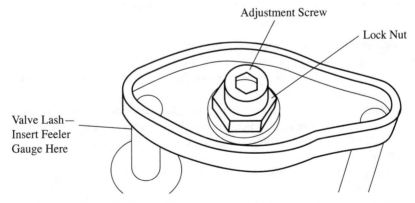

FIG. 7-6 *Lash for pressed-steel fulcrum-mounted rocker arms is measured at the pushrod-rocker interface.*

No or barely perceptible compression

No compression makes itself known as a refusal to start accompanied by zero or nearly zero resistance to the manual or electric starter. Likely culprits are a blown head gasket or stuck-open valve or, on two-stroke engines, a scored or severely worn cylinder bore.

You can get some idea of the condition of two-stroke cylinder bores by removing the spark plug and illuminating the bore with a narrowly focused inspection light of the sort sold by Granger. Deep scores or discoloration extending into the ring travel area are evidence that the engine has seen its last days.

Compression test

If the engine has an automatic compression release, disable it by removing the pushrod on the associated valve, usually the exhaust. Set manual compression releases in the "run" position.

Auto parts stores carry a variety of compression gauges for gasoline engines, but diesel gauges are another story. Available gauges are intended for industrial or automotive engines. In the absence of anything better, mount a 1000-psi pressure gauge on a gutted injector or fabricate the properly sized adapter from scratch (Fig. 7-7). A resettable check valve, scavenged from a commercial compression gauge, will stabilize gauge readings.

Bear in mind that cranking compression readings are less than definitive. Readings depend on the gauge, the length of the connecting hose, and on how fast the engine turns over. Ideally, the test should be run when the engine is new and periodically afterwards to obtain comparative results.

To test cranking compression:

1. Warm the engine to operating temperature.
2. Disengage the compression release if fitted.
3. Install the gauge in the spark plug or injector port. Ground the spark-plug lead to the engine block.
4. Make sure that the engine cannot start during the test and blow out the gauge. Disable the ignition on multicylinder gasoline engines, and shut off the fuel supply on diesel engines, regardless of the number of cylinders.
5. Open the throttle full wide.
6. Crank the engine.
7. We are looking for something on the high side of 100 psi for gasoline engines and four times that for diesels. As a point of interest, two-stroke

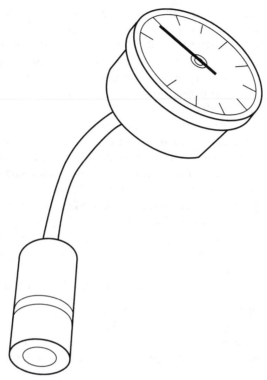

FIG. 7-7   *A diesel compression gauge with a fabricated adapter.*

engines, at least in my experience, need 90 psi to start dependably. Four-cycle spark-ignition engines sometimes can be persuaded to start with as little as 60 psi, although power will be seriously compromised.

If compression is low on a gasoline engine, squirt a little oil into the cylinder via the spark plug port. Pull the engine through several times, and retest. If the reading increases, the rings are worn, stuck, or broken. No improvement in compression means a head gasket or valve leak.

WARNING: Do not add oil to diesel cylinders. To do so is a proscription for an engine start and a gauge explosion.

Leak-down test

Introducing compressed air into the cylinder and noting how quickly pressure dissipates is the best method of determining the condition of the cylinder and, by extension, the overall health of the engine. And the test is universal, with application for all engines regardless of type.

Pressure enters the cylinder at the spark plug or injector port and leaves through leakage past the piston and/or valves. There is no agreed-on standard for these testers; leakage that one gauge reports as 10 percent, another might call 20 percent. Bore diameter also affects the leak-down rate.

In other words, cylinder leak-down tests, like cranking compression tests, have a comparative value. One should test the engine after break-in to establish a baseline reading. Only then will subsequent tests be entirely meaningful. Leakages of more than twice the baseline figure are cause for concern.

An OCT professional-quality leak tester costs around $100, and one can pay nearly five times that much for a Moroso unit, favored by auto racers. The design presented here is based on one by Mike Nixon, one of the most knowledgeable techs in the business. His original single-gauge design can be seen at www.motorcycleproject.com/motorcycle/text/leakdown.html. I added a second gauge upstream of the restrictor as a check on the regulator's ability to hold pressure (Fig. 7-8).

Parts needed, together with Granger catalog prices, include:

- Two 2-in.-diameter, 100-psi pressure gauges ($10.22 each)
- One Speedaire pressure regulator ($22.55)
- One 1/4-in. pipe thread Speedaire quick-disconnect coupling ($16.21)
- Two 1/8-pipe thread male Speedaire quick-disconnect plugs ($1.98 each)
- One 1/8-male by 1/4-female pipe thread brass reducer bushing ($5.88)

Plumbing components can be purchased locally.

The adapter—the part that makes up to the spark plug boss—consists of an eviscerated spark plug. Shatter the ceramic insulator, grind off the retain-

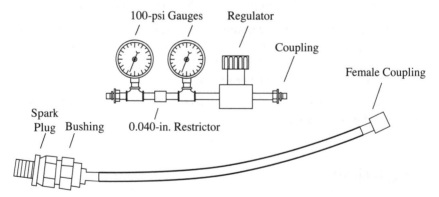

FIG. 7-8 *A homemade leak-down tester gives better results than most commercial instruments.*

ing rim, and drive out what remains of the ceramic with a punch. Cut threads on the shell ID with a 1/4-pipe tap. For diesel engines, use the adapter described earlier.

The restrictor is made by filling a 1/8-in. pipe coupling with epoxy and drilling the 0.040-in. hole with a No. 60 bit. The 40 thousandths orifice diameter follows custom. Readers might wish to experiment, although 0.040 in. gives good results on engines with bore diameters as small as 1-1/2 in.

Leak-down test—four-cycle engine

Any pressure above 15 or 20 psi can be used, although 100 psi makes calculating the percentage of leak-down easy: if the gauge nearest the regulator reads 100 psi and the second gauge reads 90 psi, leak-down is 10 psi, or 10 percent.

1. Bring the piston up to TDC on the compression stroke. Disable the compression release by removing the pushrod for the associated valve.
2. Listen for leaks: air escaping from the breather and dip stick represents blowby past the rings; leaks from the valves show up at the air filter or exhaust outlet. Note the percentage of leakage.

Leak-down test—two-cycle engine

Leak-down tests for two-cycle engines detect leaks in the intake tract rather than cylinder-bore wear. Remove the muffler and carburetor, and fabricate gasketed plates to seal off the intake and exhaust ports. Apply low pressure—*not to exceed 8 psi* through the spark plug boss. Brush soapy water on every mating surface, with special attention to the seal at the power-takeoff end of the crankshaft. If the engine has been running lean, remove the flywheel for access to the ignition-side seal. Bubbles indicate the presence of leaks.

Blowby gauge

Pistons do not seal perfectly: some combustion gases pass by the rings and enter the crankcase. In two-stroke engines, these gases pass into the combustion chamber and are burned. Four-stroke engines vent blowby through a breather and into the carburetor. The breather consists of an oil trap and an automatic valve. The valve opens during the downstroke of the piston and closes during the upstroke so that the crankcase maintains a slight vacuum. If the breather become oil-logged or the valve sticks, crankcase pressure goes positive. Oil consumption increases, gaskets weep, and crankshaft seals may fail.

A vacuum gauge connected via a rubber stopper to the oil filler tube will establish the presence of crankcase vacuum and, by extension, a functional breather. A water manometer enables more precise measurements of crankcase pressure and gives serious tuners the opportunity to measure natural gas and propane delivery pressure and the pressure drop across air filters. A manometer can also be used to measure two-stroke crankshaft pressure. I wouldn't be without one. As with other gauges, meaningful results require a baseline reading taken before problems develop. What follows is based on a Kohler design.

1. Secure a 6-ft-long, ½-in.-diameter clear plastic tube to a 10 × 30-in. piece of ½-in. plywood (Fig. 7-9). The U-shaped bend in the tube should be gradual.
2. Measure off a series of twenty-four 1-in. increments on the plywood, and number the increments from the center out beginning with zero, as

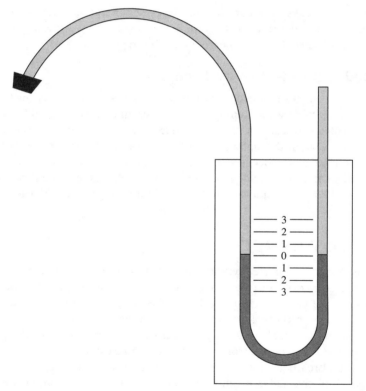

FIG. 7-9 *A manometer is a versatile tool with many applications, such as, for example, checking propane pressure on gas-fueled gensets.*

shown. Drill a hole for the tubing in a rubber stopper sized to fit the engine's oil filler tube. Home Depot can provide the stopper.

3. Position a clamp near the stopper to prevent the engine from drawing water in during cranking.
4. Fill the tube to the 0 level with water colored by vegetable dye.

To use the manometer:

1. Insert the stopper in the oil-filler hole with the clamp closed.
2. Start the engine, and run it up to full governed speed.
3. Release the clamp, and note the water level. The rise in water level on the vacuum, or engine, side of the gauge indicates crankcase vacuum in inches of water. If crankcase vacuum is less than the previously established baseline:
   - Disassemble and clean the breather.
   - Look for oil traces that indicate leaking crankcase seals or gaskets.
   - Perform a leak-down or compression test to detect leaking valves or a blown head gasket.
   - Clean the spark arrestor, and check for a restriction in the muffler.

Cylinder head casting

Removing the cylinder head opens the bore and valves for inspection and often can be done with the engine in place on the generator frame. It's good practice to loosen the head from the center bolts outward, working in a criss-cross sequence.

Carbon streaks outboard of the head gasket mean that the gasket has leaked. Scrap off the gasket residue with a single-edged razor blade. Work slowly, being careful not to scratch the base metal. Carbon deposits respond to a dull scraper and a wire wheel.

You may want to use a piece of plate glass to check the flatness of the gasket surface (Fig. 7-10). As a rule of thumb, warp should not exceed 0.002 in.

To resurface the head casting:

1. Tape a piece of medium-grit (240 or 280) wet or dry silicon-carbide paper to a piece of plate glass or a machine work table. Cutting oil speeds the work.
2. Grasping the casting near the center, move it in a figure-eight pattern over the abrasive.
3. When no more low spots are seen, finish the job with 400-grit paper.
4. Clean the head thoroughly to remove all traces of abrasive.

Install the head with a new gasket and without the use of sealant (Fig. 7-11). Factory manuals include the torque sequence, which helps to ensure

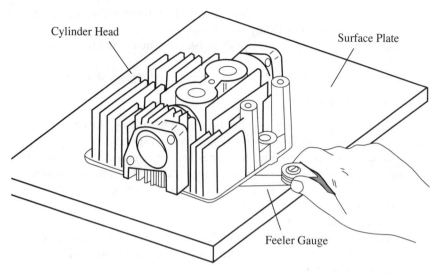

FIG. 7-10   *A piece of plate glass or a machine work table can be substituted for the surface plate shown here.*

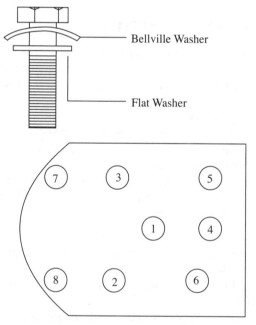

FIG. 7-11   *A DIY, no-guarantee torque sequence begins at the center of the casting and works outward. Note that Bellville washers go with their crowns up, toward the bolt head, and that flat washers are special items, not interchangeable with the hardware-store product.*

that the head goes down flat. This information will be available from the dealer and can be found on the Internet for popular engines. As a very general rule, cylinder heads and other castings should be torqued down in a crisscross fashion from the center bolts outward. In any event, approach final torque incrementally—25 percent of the torque limit, then 50, 75, and 100 percent.

Head gaskets for diesel engines come in several thicknesses as a means of maintaining the original deck height or the spatial relationship between the piston crown and the sealing surface at the top of the cylinder. Selecting the appropriate gasket is a complex procedure best accomplished with reference to the factory manual.

Valves

Figure 7-12 illustrates valve nomenclature. As a preliminary step and as a test after servicing, spray WD-40 on the valve faces, and apply compressed air to

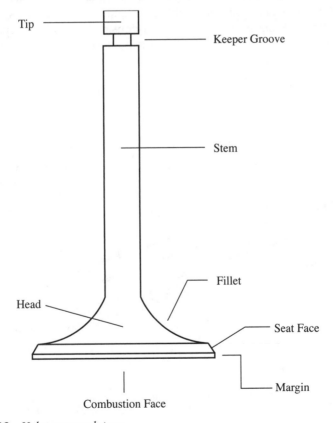

FIG. 7-12 *Valve nomenclature.*

the inlet and exhaust ports. A rag wrapped around the air nozzle will help to contain the pressure. Bubbles indicate leaks. Extracting the valves on side-valve engines requires a spring compressor, such as the one shown in Fig. 7-13. Cover the oil-return port in the floor of the valve chamber so that a loose keeper does not find its way into the crankcase (Fig. 7-14).

Overhead valves are more accessible. Place the head on a wood block to protect the gasket surface. The weak valve springs on some engines can be compressed with finger force. If the springs are too stout or the fingers too weak, position a deep-well wrench socket over the valve head. A sharp rap with a hammer compresses the spring far enough to release the keepers. Assembly requires a spring compressor.

The contact area on the valve face should be dull bright, without interruptions, pitting, or grooving. When in doubt, dab a small amount of marking compound on the seat, and rotate the valve on its seat. Prussian Blue, available at machinist supply houses, is nasty stuff to work with because it transfers to hands, clothes, and tools. This quality also makes it an excellent tell-tale. The bluing trace should be continuous, unbroken, and centered on the valve face.

Older Briggs & Stratton, Kohler, and Wisconsin engines have replaceable valve seats, which is both a blessing from the point of view of reparability and

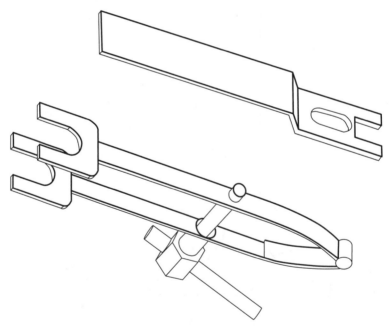

FIG. 7-13 *An overhead valve compressor (above) and the more complex tool required for side valves. OHV tools vary with the engine model but can be easily fabricated.*

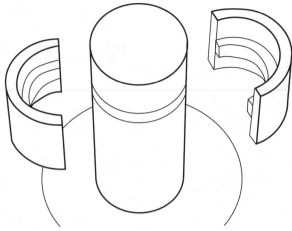

FIG. 7-14 *Valve keepers are wedges that are held in place by the valve spring. Check that the keepers fully seat in their grooves on installation. Should an OHV keeper come loose, the valve falls into the chamber and collides with the piston. You can imagine the effects.*

a curse when the seats come loose. Overhead valve engines have cast-in seats that rarely, if ever, come loose. But should a seat crack or otherwise fail, the head must be replaced unless you can find a really expert machinist.

Back in the days of flat-head Fords, Bull Durham, and Western Auto, mechanics routinely lapped valves with Clover Leaf grinding compound. Unfortunately, much of the effort was in vain. The metal-to-metal seal created by lapping disappears when the engine starts and the parts undergo thermal expansion (Fig. 7-15). Even so, lapping does have one application.

Automotive machine shops grind valve seats, an operation that involves a considerable investment in tooling. Small shops and DIY mechanics are more likely to use a Nu-Way or equivalent refacing tool. These hand-operated tools leave a wavy surface that can be smoothed by judicious lapping.

Most valve faces are ground at 45 or 46 degrees. Briggs & Stratton specifies 30 degrees for some intake valves to improve cylinder filling. The primary seat angle matches the valve-face angle and is flanked by a shallow 30-degree entry angle and a 60-degree undercut, as shown in Fig. 7-16. The combination of angles determines the seat width, which is also a matter of specification, as is the thickness of the valve-face margin. In addition, diesel engines are particular about the distance the closed valve extends into the combustion chamber. And no engine responds well to buried valves that result from over-enthusiastic seat grinding.

Do not trust a machinist to back-engineer specifications from the parts; supply him or her with the documentation that can be had from the genset dealer or from websites listed at the back of this book.

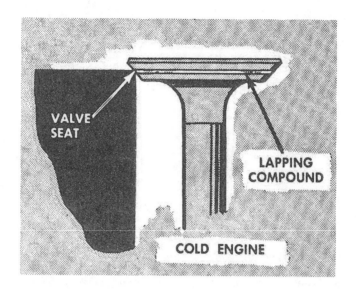

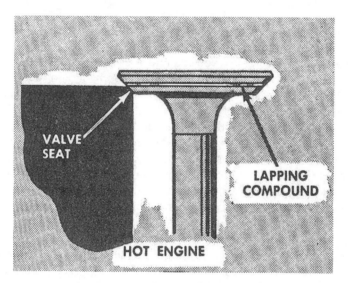

FIG. 7-15 *Lapping valves concaves the valve face because valve steel is much softer than Stellite and similar seat materials. When the engine warms and the seat moves as a result of thermal expansion, the valve leaks. Thermal expansion also can upset valve seating on new or low-hour engines.*

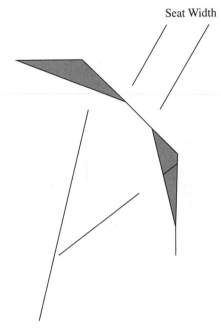

Seat Width

Remove Metal Here

FIG. 7-16 *The seat angle matches or slightly interferes with the valve face angle. Entry and exit undercuts determine the seat width and the unit pressure applied to the valve face. Most seats have a 30-degree entry angle, followed by a 45-degree contact angle and a 60-degree undercut. But these angles are not set in stone. After months of intensive development, Harley-Davidson engineers showed their new model 103 and 110 cylinder heads to Joe Mondello, a legendary tuner. The engineers were taken aback when Joe tweaked the seat angles to boost flow by 52 ft³/m.*

Valve guides

Guides bell-mouth and taper due to side forces generated by the rocker arms. The workaday wear limit is 0.004 in. (Fig. 7-17). Mechanics, experienced ones anyway, judge guide condition by the amount of wiggle the valve exhibits in the full-open position. A more accurate assessment involves multiple measurements with a ball gauge.

Worn guides present three options:

- A machinist may be able to recondition the guides with sleeves, which cost less and are easier to install than the guides themselves.
- Some OEMs supply valves with oversized stems to allow the existing guides to be retained.
- The guides can be replaced, which is the traditional approach.

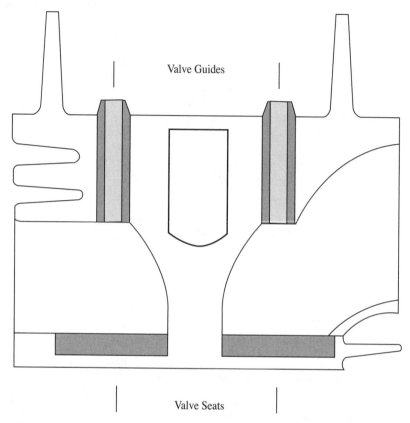

Valve Guides

Valve Seats

FIG. 7-17 *Valve-guide materials give insight into the build quality of the engine. Cast iron is the base choice, followed by manganese or phosphor bronze. Heavy diesel and some motorcycle engines now use powdered-metal guides, such as Federal-Mogul PMF-10F, a nickel-steel alloy fused with solid lubricants.*

If you opt to replace the guides yourself, begin by measuring the installed depth—the distance from the valve seats to the tops of the original guides. Side-valve guides drive down and out and install from above. How overhead valve guides are accessed depends on the head. There is always a way to reverse the factory installation process. Splintering the guide with a chisel is sometimes necessary on side-valve engines with small valve chambers.

Replacement guides are often pressed into place, although careful mechanics chill the guides on aluminum engines to prevent damage to the base metal. Placing the guides in a freezer and warming the casting to 160 to 180°F permits the guides to slip into place without violence. Should one stick, place a plastic or soft-wood buffer on top of the guide, and tap it home. In most cases the installed guide will have to be reamed to size, an operation that calls for a special factory tool or a good-quality adjustable reamer. Cheap reamers are a contraction in terms.

Although factory manuals warn that guides should be installed dry in order to assist heat transfer, in an emergency you can roughen a loose guide with a rasp and anchor it with Loctite 271.

Valve springs

It is always good practice to replace the valve springs in the course of cylinder head work. Springs must be replaced if shorter than specified or if the spring does not stand vertical. Many intake- and exhaust-side springs interchange; others differ in terms of stiffness. A few have more closely wound dampener coils at their stationary ends (Fig. 7-18) Yanmar identifies the dampened end with paint.

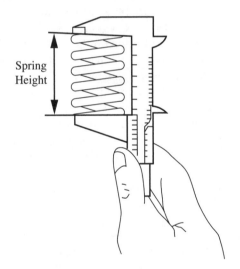

FIG. 7-18 *Serviceable springs stand square and meet factory specs for free length.* Kohler Lombardini

Pushrods

Pushrods are often used to transfer oil to the rocker assembly. Make sure the passageways, which tend to collect slit, are clean. OEMs are strict about pushrod straightness since any deviation creates a moment that invites further bending. Yanmar, for example, limits allowable bend to 0.002 in., checked on plate glass with a feeler gauge. That said, I have used slightly bent pushrods without immediate problems.

Pistons

Most engines give access to the piston assembly by removing the oil pan or side cover; some vintage two-strokes have detachable cylinder jugs. Whatever the arrangement, the first order of business is to "read" the carbon.

Soft, furry carbon that wipes off on the fingers means that the engine has run rich, often the result of a restrictive air filter. By the same token, the near absence of carbon accompanied by yellowish or white discolorations is evidence of higher-than-normal combustion temperatures. Suspect air leaks or a clogged carburetor. Stratified deposits that range from gummy to hard-black flakes indicate oil burning. Dark, dry, and hard carbon is the ideal.

Note that the underside of the piston crown is a far better indicator of combustion temperature than spark plugs. Stains should be dark brown, verging on black: the lighter the shade, the higher is the combustion temperature.

The next step is to scrape the carbon off the piston top using a dull tool to protect the finish. Most pistons will be marked with an arrow or other symbol. Note the relationship between this mark and the flywheel or another prominent feature as an installation guide.

Examine the piston thrust faces for wear or, more exactly, the quality of wear. Deep scores condemn the piston and almost always damage the bore. A satiny surface that has the texture of a cat's tongue means imbedded abrasives that usually enter through a defective air filter. Other sources of contaminates are leaking gaskets at the air filter or carburetor flange, loose carburetor mounting bolts, or a worn carburetor throttle shaft. Whatever the cause, contaminated pistons must be replaced and the cylinder bore machined oversize.

With the piston at TDC, try to move it up and down. One or two thousandths inch play in the connecting-rod–crankpin bearing is normal, as is a smidgeon of side-to-side wobble. Greater play suggests excessive wear.

The final test is to insert a feeler gauge between the compression rings— the upper two rings on four-cycle engines or all the rings on two-cycle engines—and the sides of the ring grooves. The wear limit ranges from 0.006 to 0.002 in. depending on engine make and model (Fig. 7-19).

To remove the piston, remove the pin locks, called circlips, and force the wrist pin out the tool shown in Fig. 7-20. Heat will also loosen the pin. The least desirable technique is to drive the pin out with a flat punch.

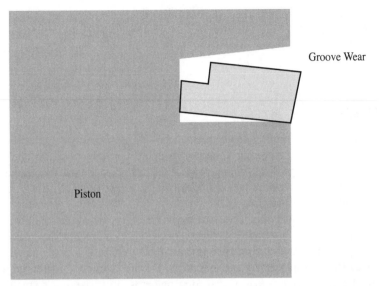

FIG. 7-19 *The rings—especially the upper compression ring—batter their grooves as the piston reverses direction twice each stroke. The upper faces of the grooves wear at an angle, giving the rings freedom to twist and fatigue. For durability, ring grooves on industrial engines are often lined with steel.*

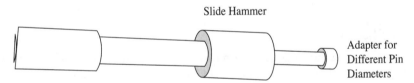

FIG. 7-20 *Piston-pin pullers are replacing older-style tools that used a threaded rod to push the pin out.*

Inspect the piston pin for scores and heat discoloration. Normally, one sees very little wear on the associated bearing surfaces. Mount the piston with new circlips for the insurance value. Should a circlip fail or not be seated in its groove, the piston pin will walk into the bore, destroying the engine.

CAUTION: Pistons can be inadvertently installed 180° out of their original position. Should this happen, the offset pin will accelerate wear on the leading thrust face.

New rings require that the cylinder bore be scuffed to assist break-in. See the "Glaze breaking" section later in this chapter for details. If the original rings are reused, the bore merely needs to be wiped down and oiled with a wet rag or sponge.

Rings

Modern four-cycle engines have three rings—a compression ring at the top, a second compression ring or a scraper ring in the middle position, and an oil ring in the bottom groove. Two-cycle engines have two identical compression rings fixed in their grooves by pins. Were the pins free to rotate, the ring ends would foul on the ports and break.

Prudent mechanics check the end gap of each replacement ring by inserting the ring in the bore with the piston crown as a pilot to keep the ring square. The gap between ring ends is then measured with a feeler gauge. Most OEMs call for 0.0015 in. of gap per inch of bore diameter. Too large a gap means that the bore is worn or that the ring set is undersized. Too small a gap means the rings will touch ends and shatter once operating temperature is reached. Correct by filing the ring ends square with a file.

Some DIY mechanics install 0.010-in. or 0.25-mm oversized rings to compensate for wear in standard-sized bores. Whatever the wisdom of this approach—the friction induced by oversized rings costs fuel and further bore wear—it is vitally necessary to file the gaps, which will be smaller than normal.

Installation

When replacing the rings on a used piston, cleaning the ring grooves is critical. The tools sold for this purpose work all too well in that they shave metal. At least, that's been my experience. One ends by breaking an old ring, mounting it in a file handle, and scraping.

CAUTION: Ring edges are as sharp as glass.

Identify each ring, and lay the rings out in the order of installation, with their top faces up.

Position the bottom ring in an expander tool such as the one shown in Fig. 7-22. The ring must lay flat in the tool—rings snap when twisted. Expand the ring just enough to clear the piston, and install it in its groove. Do the same for the remaining rings.

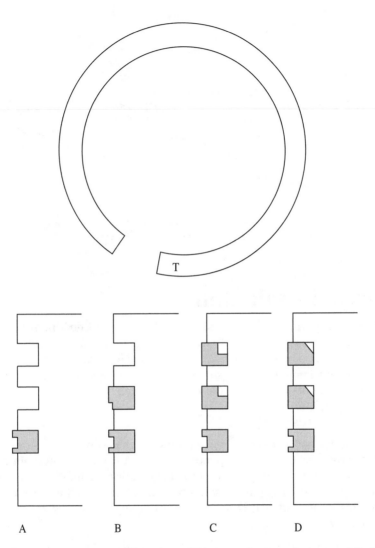

FIG. 7-21 *Four-cycle pistons have three rings—an oil ring at the bottom (A). Some use a scrapper ring (B), but the upper two rings on most pistons are compression rings (C, D). Scrapper and compression rings must be installed in the correct orientation with the marked side up toward the piston crown.*

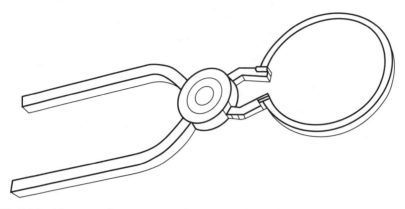

FIG. 7-22 *More complex ring expanders are available, but the simple tool shown here is easier to use. Just make sure that the ring seats flat.*

Piston installation

At this point in the story, the bore should be honed (if new rings are used) and cleaned as described under "Glaze breaking" later in this chapter. The ring compressor tool shown in Fig. 7-23 must be used on engines with detachable cylinders and is preferred for all engines. The old-style compressor, which resembles a blood-pressure cuff, is a ring breaker.

Integral bores

Most cylinder bores are of a piece with the block casting. The piston installs with the rod shank suspended, pendulum-fashion, from the piston pin. As the piston goes home, the upper half of the big end bearing must be guided into position over the crank pin. The rod cap is then installed with cap and shank match marks aligned as described in the "Connecting rods" section later in this chapter.

- Cover the rod studs with short pieces of fuel line to prevent damage to the crank pin.
- Lubricate all friction surfaces, with special attention to the big-end rod bearings, crank pin, the underside of the piston, and the ring stack. Wipe down the cylinder with an oil-soaked sponge or rag.
- Rotate the crankshaft to its lowest position.
- Stagger the ring gaps on the pistons of four-cycle engines, and make certain that two-cycle ring ends butt against their pegs.

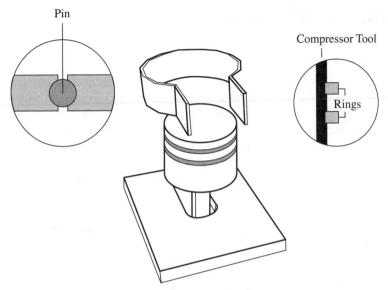

Pin

Compressor Tool

Rings

FIG. 7-23 *Rings must seat in their grooves and, on two-cycle pistons, butt against their anti-rotation pegs. A slotted wood block, used on engines with detachable cylinders, prevents the rod from flopping about.*

- Compress the rings, and force the piston out of the tool and into the bore. The force can be applied with a hammer handle. If the piston hangs, a ring has escaped the tool, or the rod has fouled against the crankshaft.
- Once the rod is over the crank pin, remove the fuel-line bumpers, flood the bearing and crankpin with motor oil, and pull the rod down and into full bearing contact.
- Install the rod cap, and torque it down to specification.
- Turn the crankshaft through several rotations. The crank will offer more resistance at midstroke, but it should not bind. The whisper made by new rings against a honed bore is a kind of music.

Detachable cylinders

Some two-cycle engines have detachable cylinders, or jugs. Lubricate the cylinder bore, piston pin, and ring stack. Things go easier if the piston assembly is stabilized with a wooden fork, as shown in Fig. 7-23. A bevel present on the underside of many cylinder bores compresses the rings as the jug is lowered over the piston. Engines without that feature require a clamp-type compressor or strong, cut-proof fingers.

Connecting rods

Construction

Connecting rods are heavily stressed components that must withstand severe bending and axial forces. On most genset engines, the rod also doubles as the big-end bearing, which is less than ideal. Upgraded rods, while intended for racing applications, are not necessarily a bad investment. Figure 7-24 illustrates one such rod for Briggs & Stratton engines.

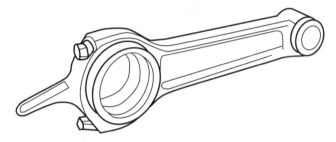

FIG. 7-24　*ARC high-performance connecting rod with replaceable bearing inserts, gently flared changes in cross section, and Allen-head rod bolts is a practical option for many Briggs & Stratton engines. Like most small-engine rods, this example has an extension on the cap, known as a dipper, to splash oil about the crankcase.*
E.C. Carburetors

Orientation

Pistons and connecting rods have a definite orientation. The piston crown is stamped with an arrow or other symbol. The connecting-rod shank and cap are also marked (Fig. 7-25). The rod match marks must align with each other on assembly, and the piston mark must be returned to its initial orientation with the rod marks and the flywheel end of the crankshaft. This sounds more complicated than it is.

Bearings

Bushings and other plain bearings skate on an oil film, much like a hydroplane. Once the engine starts, there is no metal-to-metal contact. By the same token, loss of oil flow means almost instant destruction. Fortunately, modern gensets short out the ignition if the oil level or pressure falls dangerously low.

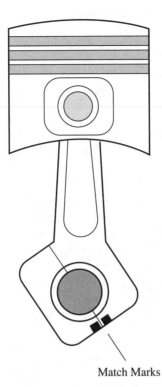

Match Marks

FIG. 7-25 *Rod shanks and caps have a matching pairs of marks that must be aligned as shown.*

Connecting rods that form the bearing surface cannot be reworked. Filing the caps merely ovals the bearing, reducing its contact area. Some manufacturers supply 0.010-in. or metric-equivalent undersized rods to fit reground crankshafts. But have the replacement in hand before commissioning machine work.

More durable engines have insert-type big-end bearings, similar to those used on automobile crankshafts. These bearings come in one and sometimes two oversizes to enable worn crankshafts to be reground.

Insert bearings lock into place with a tab and are further secured by friction (Fig. 7-26). That is, the diameter of the bearing shells is a few thousandths of an inch larger than the diameter of the saddle. When the cap is tightened, the bearings undergo a controlled degree of crush.

Two-cycle engines, at least those designed with longevity in mind, use loose needle bearings on the big end and a cartridge needle bearing on the

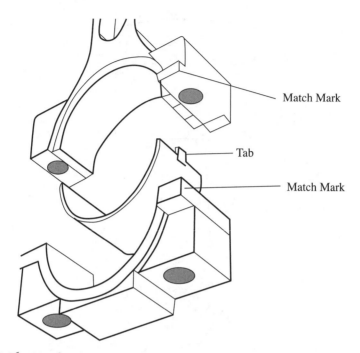

Match Mark

Tab

Match Mark

FIG. 7-26   *Most bearing wear occurs on the lower shell, which absorbs the loads developed during the expansion stroke. Diameters—STD, 0.010, and sometimes 0.020 for the inch standard—are printed on the back of the shell. Upper and lower shells may differ in that one has an oil port. Tighten the cap evenly in three or more increments to the specified torque limit.*

small end. Rolling, or antifriction, bearings survive on a whiff of oil vapor. Aluminum rods for these applications have steel bearing races.

There is no positive way to detect wear in antifriction (i.e., needle, roller, or ball) bearings other than by visual signs such as discoloration or skid marks on the races. When in doubt, replace the bearings.

Loose, uncaged needle bearings run and hide during disassembly. As a general rule, as many needles are used that fit. In other words, if there is space for another needle, one is missing. Do not mix old and new needles; the worn elements will not share the load. Secure the bearings for assembly with low-temperature assembly grease. Old-timers used beeswax.

Small-end rod bearings rarely give trouble unless the piston has overheated, a condition that will be indicated by light brown or white deposits on the underside of the crown. Small-end bushings can be purchased at a bearing supply house and require reaming after installation.

Most plain-bearing connecting rods—those that use the rod itself or an insert for the big-end bearing—require 0.0015- to 0.0020-in. running clearance. Crankshaft taper and out-of-round should be no more than 0.0010 in. But don't take my word for it: verify these supercritical specifications with the factory service manual.

There are at least three ways of determining big-end bearing clearance:

- Use the piston as a rough indicator of bearing wear. Bring the piston to top dead center. Slowly turn the flywheel a few degrees on one side of TDC and then in the other. Note the effect on the piston. Some disconnect between flywheel rotation and piston response is normal. But if you can move the flywheel a sixteenth of an inch or more with the piston stationary, the big-end bearing has failed. This somewhat homely technique works for both plain and antifriction big-end bearings.
- Measure the crank pin and big-end diameters with inside and outside micrometers.
- Measure plain-bearing clearance with Plastigage.

Micrometers

Professional-quality work requires outside and inside micrometers. Figures 7-27 and 7-28 illustrate how metric micrometers are read. Inch-standard micrometers work on a similar principle but with different increments.

Although less precise than micrometers, calibers are a useful addition to any mechanic's tool collection (Fig. 7-29). Double jaws enable both inside and outside measurements. While most technicians prefer digital or mechanical dial calipers, Starrett or Mitutoyo vernier calipers can sometimes be purchased for a few dollars at garage sales and never drift out of calibration.

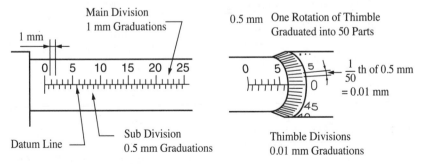

FIG. 7-27 *A metric micrometer.*

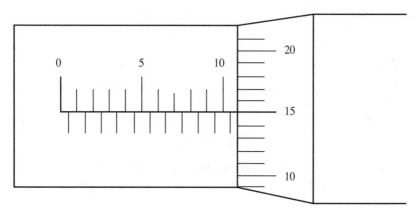

FIG. 7-28 *Ten millimeters on the datum scale, 0.5 mm on the subdivision scale, and 0.15 mm on the thimble give 10 mm + 0.50 mm + 0.15 mm = 10.65 mm.*

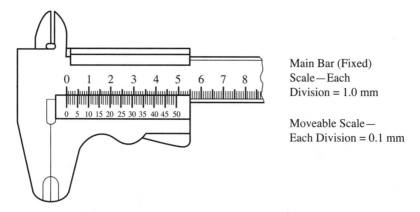

FIG. 7-29 *Read a vernier from the main scale. The numbered divisions each represent 10 mm. What remains on the right of that large number is determined by the vernier division that aligns with a main-scale division. In this example, vernier 6 indexes with main scale 25. The caliper reads 25 + 0.6 = 25.6 mm.*

Plastigage

Fel-Pro Plastigage comes in two ranges—green for 0.001 to 0.003 in. (0.025 to 0.075 mm) and red for 0.002 to 0.006 in. (0.050 to 0.150 mm). Crankshaft taper and out-of-round should be less than 0.001 in. (0.025 mm).

To use:

1. Turn the crankshaft to position the rod at bottom dead center (BDC).
2. Remove the rod cap.
3. Wipe the oil off the rod and crank pin.
4. Break off a piece of gauge wire and lay it lengthwise along the crank pin.
5. Replace the rod cap, being careful not to move the crankshaft.
6. Undo the cap, and compare the width of the wire with the scale printed on the envelope. The average width represents bearing clearance.
7. Scrape off the wire fragments, and repeat the test, this time with two pieces of gauge wire laid athwart the crank pin. Variations in the width of the two wires indicate the amount of taper.

CAUTION: Plastigage that crumples when handled has exceeded its six-month shelf life.

Failure analysis

Scores on the big-end rod bearing that may be accompanied by a dull, satiny appearance indicate the presence of abrasives in the oil. Check the air filter for loose fasteners and other leak sources. Paper elements should not leak bright light when held up to the sun. Abrasives also can enter by way of a worn carburetor throttle shaft or loose carburetor mounting bolts.

The worst catastrophe that can befall an engine is for the connecting rod to part. The piston, acting like a bullet, descends to bring the severed rod shank into the path of the spinning crankshaft. When the parts collide, the rod is often thrown outward with enough force to penetrate the crankcase.

The causes of a thrown rod center on the big-end bearing. Overheating discolors tin-plated rods and may leave aluminum smears on the crank pin as the protective oil film vaporizes. When the bearing seizes, the rod shank breaks. Loss of lubrication has a similar effect.

Insufficient torque on the rod cap pounds out the bearing and, in extreme cases, fragments the cap. Some connecting rods depend on lock washers or locking tabs—which should be renewed whenever disturbed—to prevent backing off. Others rely entirely on assembly torque.

Fatigue failure usually occurs at the midpoint of the shank. You may see evidence of crystallization at the break. Overloading, as when the generator stalls under excessive loads, also breaks the rod somewhere near its midpoint. This sort of failure can sometimes be recognized by the preliminary bend before the rod gives way.

Rod assembly

Oil both halves of the rod bearings and the crank pin liberally, making certain that all surfaces are wetted. Install the cap with match marks aligned, and torque to specification in three or more increments. Once installed, the rod should rotate freely and move from side to side on the bearing without protest.

Cylinders

Glaze breaking

For new rings to seat, the protective glaze that forms on cylinder bores must be roughened with a flex hone, also known as *ball, bottle*, or *dingle-berry hone*. The process is a messy one, spilling aluminum-silicon particles everywhere. Consequently, the crankshaft and other engine internals must be removed prior to honing.

Tape the old head gasket in place to protect the sealing surface, and run the hone up and down the cylinder to produce a cross-hatch pattern that provides the short-term abrasion new rings require. Lightly does it since the goal is to abrade and not remove metal.

Clean the bore with hot water and detergent to float abrasive particles out of the metal pores. Dry with ordinary paper towels, alternatively soaping and drying until the towels no longer discolor. Much scrubbing and rolls of towels later, the bore can be considered clean and ready to be sprayed with motor oil or WD-40.

Inspection

Cast-iron blocks are quite soft and over time develop a perceptible step, or ridge, at the upper extreme of ring travel. The depth of the ridge provides some idea of bore wear but says nothing about out-of-round or taper. The sleeves used on aluminum-block engines do not wear in such a dramatic fashion but are very prone to distortion, which can occur in new engines as casting stresses normalize.

The combined effects of bore distortion and wear compromise the ability of the rings to hold pressure, which is the prerequisite for power production.

Abrasives entering by way of a leaking air filter or leaks around the carburetor throttle imbed themselves in the bore metal. Two-cycle cylinder bores develop deep scores from carbon particles that fail to exit the exhaust port or flake off from the combustion chamber. If overheated or poorly lubricated, these engines weld their pistons to the bores, depositing aluminum splatter that can be removed with muriatic acid. Blowby discolors the bore in all engines, and broken rings leave deep, fingernail-hanging grooves. Should a circlip fail, the piston pin "walks" into the cylinder wall, destroying the engine.

Most mechanics judge bore oversize by piston fit, a technique that requires experience. Another rough indication of bore condition can be had with a piston ring. Insert the ring into the bore using the backside of a piston as a pilot to keep the ring square. Variations in the ring gap, measured with a feeler gauge at different bore depths, say something about wear and distortion.

Ideally, one would use a cylinder bore gauge, but reliable versions of the tool cost several hundred dollars, and require periodic calibration. Telescoping gauges are an inexpensive, although a skill-intensive, alternative (Fig. 7-30). Rock the gauge back and forth to be sure that it's held perpendicular to the bore.

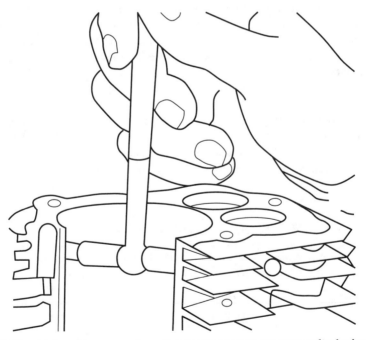

FIG. 7-30 *A telescoping gauge is an inexpensive way to measure cylinder bore diameter, taper, and out-of-round. The spring-loaded legs ensure the correct contact force. Once the measurement is made, the legs are locked and their extension measured with an outside micrometer or calipers.*

Suggested bore wear limits cluster around 0.004 in. (0.10 mm). For example, Cummins puts the limit at 0.12 mm for its Honda-powered gensets. Yanmar sets the wear limit at 0.10 mm; some Briggs & Stratton OHV models have stricter limits. While adherence to factory specs is the high road, the real test of cylinder bore integrity is the ability of the engine to maintain 3600 rpm under full rated load. An engine that does this does not need reboring. A worn-out cylinder presents three options—purchase a new block casting or have a machinist rebore or sleeve the existing block.

Boring

Cast-iron side-valve engines can be overbored 0.125 in. and sometimes more. Aluminum engines do not have the reserves of metal to allow oversizes greater than 0.020 in. (0.50 mm), which is the largest piston normally available. The machinist uses a boring bar for the initial cuts to within 0.003 in. of the final size, which is achieved by honing.

Piston sizes come in 0.010-in. (or 0.25-mm) increments. As a point of interest, the relationship between the two standards is shown in Table 7-3.

Briggs & Stratton and several other manufacturers size replacement pistons relative to the original piston diameter. A Briggs 0.030-in. piston is thirty thousandths of an inch larger than the standard-sized piston. If the bore is machined 0.030 in. oversize, the replacement piston will have the correct running clearance. Confusion arises because automotive piston oversizes are based on the original bore diameter. To obtain running clearance, the machinist must bore the cylinder a few thousands of an inch larger than the piston oversize.

Piston-to-bore clearance, while a matter of factory specification, is a subject enmeshed in mythology. Clearance has no or no iron-clad relationship with bore diameter. Nor should engines be set up looser than specified as insurance against piston seizure. Sloppy, rattling pistons can and do seize.

Table 7-3
Metric and U.S. piston oversizes

Metric piston oversize	Nominal U.S. equivalent	Actual inch oversize
0.25 mm	0.010 in.	0.010 in.
0.50 mm	0.020 in.	0.019 in.
0.75 mm	0.030 in.	0.030 in.
1.00 mm	0.040 in.	0.039 in.
2.00 mm	0.080 in.	0.079 in.
3.00 mm	0.125 in.	0.118 in.

Abide by the factory specs that, for the most part, call for piston-to-bore clearances of 0.0015 to 0.0020 in. (0.038 to 0.051 mm).

Sleeving

Depending on the condition of the crankshaft and main bearings, restoring the bore with a sleeve may be the best option. Los Angles Sleeve (www.lasleeve.com) stocks more than 4000 part numbers for these thin, centrifugally cast-iron cylinders. An interference fit of 0.0040 in. for aluminum blocks and 0.0025 in. for cast-iron holds the sleeve in place. To make installation easier, many shops freeze the sleeve and apply moderate heat to the block. Cooling the sleeve in a household freezer reduces diameter by about 0.002 in. Dry ice or liquid nitrogen produces enough shrinkage for the sleeve to drop into place.

Flywheel

The flywheel nut has a standard ("left-to-loosey") thread unless you're working with an ancient Briggs horizontal-shaft engine. Secure the wheel against wrench torque with a strap wrench. If plastic fan blades make this impractical, run a length of nylon rope into the spark plug or injector boss as a piston block.

The flywheel mounts on a taper, which makes it difficult to separate from the crankshaft. Figure 7-31 shows the use of an automotive harmonic bal-

FIG. 7-31 *A harmonic balancer puller can be used to extract most flywheels.*

ancer puller, a tool that works if the flywheel hub has three bolt holes for purchase. Two-hole hubs require a special tool, which can be fabricated or purchased from a small-engine supply house.

Do not use an ordinary gear puller that grips the flywheel rim. Nor, in my opinion, should you use a knocker to shock the wheel off. The brutality of the process aside, a knocker must pit antifriction bearings and scramble flywheel magnets.

A deeply grooved key upsets ignition timing enough to make the engine hard to start; the flywheel shift that comes about from a sheared key will prevent starting. Wallowed-out keyways suggest that the flywheel nut was insufficiently torqued. Figure 7-32 shows how the flat and half-moon keys install on the crankshaft taper.

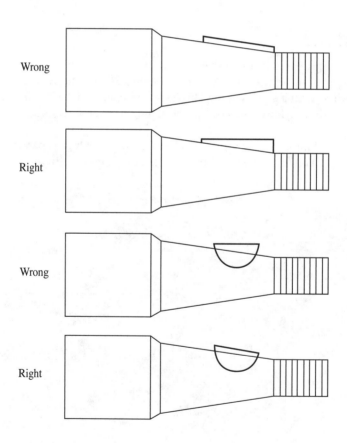

Wrong

Right

Wrong

Right

FIG. 7-32 *The correct lie of flywheel keys is a bit counterintuitive.*

When one thinks about it, a steel or Briggs aluminum key cannot be counted upon to secure the flywheel. The flywheel nut forces the tapered flywheel hub over the crankshaft taper to create the binding force. The key functions merely as the ignition timing reference and can be omitted if the keyways are aligned.

Crankcase

Nearly all four-cycle engines have horizontally split crankcases that give access to engine internals from below. The larger examples also have removable side covers.

Two-cycle engines traditionally have vertically split crankcases (Fig. 7-33). Remove the fasteners, using a hammer-impact driver if the crankcase halves are held together with Phillips-head screws. Taps with a rubber mallet usually

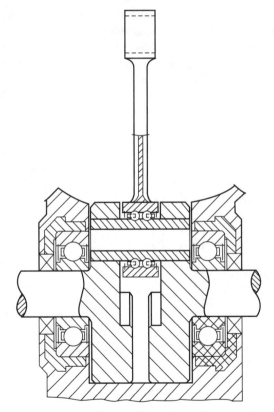

FIG. 7-33 *A two-cycle crankshaft assembly with antifriction main and crank-pin bearings (U.S. Patent No. EP0334631).*

suffice to break the gasket seal. Many of these crankcases have raised pry points to assist in separating the parts. Prying on the gasket surfaces must be avoided at all costs. The interference fit on the main bearings can be overcome with judicious use of a heat gun.

If a gasket is present between the crankcase halves and if each half houses a main bearing, the thickness of the gasket affects the amount of crankshaft end play. Most crankshafts require 0.004 to 0.005 in. of fore and aft movement to compensate for thermal expansion. If a gasket is not used, assemble the parts with a sealant such as Yamabond 4.

Crankshafts

Figure 7-34 illustrates crankshaft nomenclature.

Prior to disassembly, pull the engine through several times with the spark plug or injector removed. Focus on the central bolt hole at the crankshaft end. Wobble indicates a bent crankshaft. A more precise determination of crankshaft trueness can be made after disassembly (Fig. 7-35).

Check the keyways for side play (some of which is inevitable), and begin disassembly on four-cycle engines by turning the crankshaft to align the timing marks (Fig. 7-36). If the crankshaft gear has worn enough to obliterate its mark, rotate the crank to bring the piston to top dead center with both valves

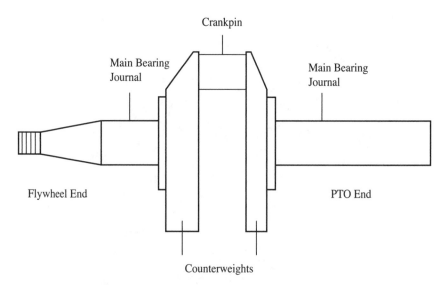

FIG. 7-34 *Crankshaft nomenclature.*

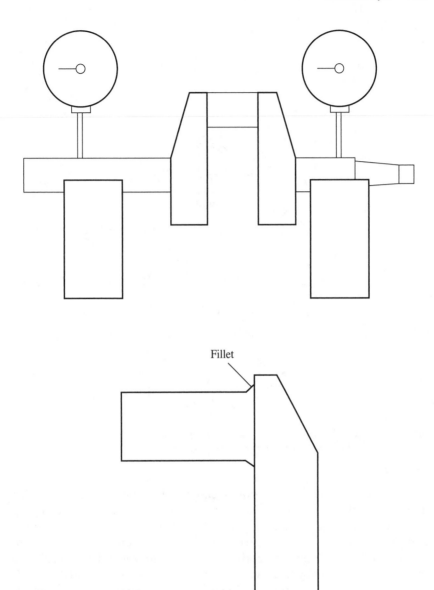

Fillet

FIG. 7-35 *A more precise way to check crankshaft straightness is to rotate the shaft on machinist's V-blocks under a pair of dial indicators. Fatigue failure nearly always originates at the fillets—the transitions between the crank pin and the webs. If you have the crank reground, make certain that the machinist leaves wide, gently radiused fillets.*

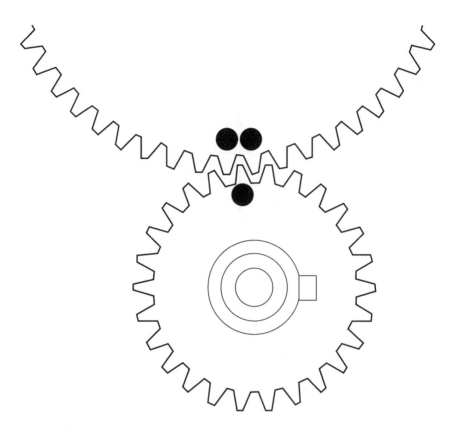

FIG. 7-36 *Timing marks usually take the form of three circular indentations. Tecumseh sometimes uses a keyway as one of the marks.*

closed. (An automatic compression release will hold one of the valves a few thousandths of an inch off its seat.) Turn the crankshaft a few degrees against the normal direction of engine rotation. The exhaust valve should crack open. Reversing the rotation a few degrees past TDC should unseat the intake valve.

Use a micrometer on the crankshaft bearing surfaces, with special attention to the crank pin, which wears far more rapidly than the main-bearing journals. As indicated previously, neither out-of-round nor taper should exceed 0.001 in. Of the two, taper is the most to be feared because it generates side forces that can drive the wrist pin past its locks and into the cylinder wall. If you have access to the data, compare the diameter of the pin with factory reject limits. Otherwise, its serviceability can be intuited from crankpin–bearing clearance.

Crankshafts are either made of ductile cast iron or, as in the case of most Honda GX, and all B & S Vanguard, and Kohler engines, of forged steel (Fig.

(a)   (b)

FIG. 7-37   *Forged (a) and cast-iron (b) crankshafts used in Dr. Fatemi's study.*

7-37). Cast-iron counterweights have a prickly surface, rough to the touch. Nonmachined surfaces on forged cranks tend to be smoother, with irregularities flattened by the hot-hammer work.

Run long and hard enough, all crankshafts crystallize and break (Fig. 7-38). According to a 2007 study by Dr. Ali Fatemi and colleagues, the increased fatigue resistance of a forged crankshaft means that it should live at least six times longer than an equivalent iron shaft (Ali Fatemi, Jonathan Williams, and Farzin Montazersadgh, "Fatigure Performance Comparison and Life Prediction of Forged Steel and Ductile Cast Iron Crankshafts," University of Toledo,

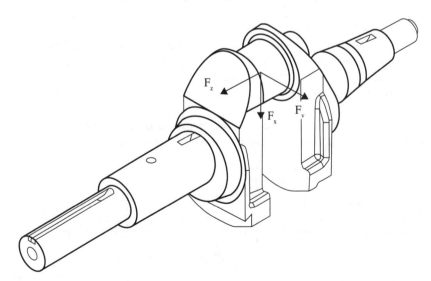

FIG. 7-38   *Three forces act on the crank pin—bending (F_x), longitudinal (F_z), and torsional (F_y). Fatigue cracks are most likely to develop at the fillet on the underside of the pin.* Dr. Ali Fatemi

Toledo, OH, 2007). Other advantages include higher tensile strength and reduced weight.

We work with what we have. But if what we have is a well-used iron crank, the few dollars an auto machinist charges for Magnifluxing can avoid the double whammy experienced when both the grid and the generator go down. If one has a good ear, the persistent ring a hammer tap on the crankshaft produces can be distinguished from the attenuated sound produced by a defective shaft. Another approach, sometimes used in the less affluent parts of the world, is to sprinkle the crank with chalk dust and tap. The dust collects in large cracks, making them visible. But neither of these expedients substitute for magnetic particle testing.

Antifriction main bearings

The better engines have ball or tapered roller-bearing mains. There is no iron-clad way of measuring wear on this type of bearing. Incipient failure is a judgment call depending upon one's perception of excessive play or roughness as the bearing is turned.

CAUTION: Spinning bearings with compressed air to watch them whirl is not recommended.

A tight interference fit between the bearing inner race and the crankshaft requires the tool shown in Figure 7-39. Bearings extracted in this manner cannot be reused.

Replacement bearings can be purchased from a bearing supply house, usually for less than the genset dealer charges. Explain to a knowledgeable person in the supply house how the bearings will be used, and take his or her suggestions about the class of fit. Engine bearings are set up looser than most industrial types.

Antifriction bearings install with the factory code numbers on the visible side of the bearing. You can have the bearings pressed on at a machine shop

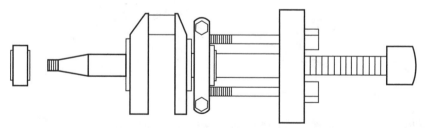

FIG. 7-39 *Bearing splitters are expensive, seldom-used tools that sometimes can be rented.*

or use a driver. The driver consists of a piece of pipe trimmed square and sized to fit the bearing inner race. No force should be applied to the outer race.

Camshafts

There's little to say about camshafts, except to reiterate the importance of verifying that the automatic compression release functions. A camshaft sees extraordinarily large forces, on par with the chamber pressures of a military rifle. A lubricant, such as Crowder ZPaste, must be used on the lobes of a new camshaft and should be used whenever the engine is opened.

Balance shafts

Briggs & Stratton, Honda, Kohler, and other top-of-the-line engines employ balancer shafts to counteract the shaking forces generated by the crankshaft (Fig. 7-40). Carefully inspect the bearings and assemble with the timing marks aligned.

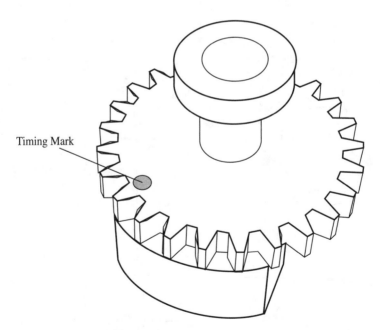

Timing Mark

FIG. 7-40 *A single balance shaft is most often encountered, although some engines have two. Briggs & Stratton once used a dummy connecting rod reciprocating at 180 degrees to the actual rod.*

Oil seals

Seals contain the crankcase oil in four-cycle engines and hold crankcase pressure in two-stroke engines. Crankshaft seals that are not part of an antifriction bearing can be replaced without dismantling the engine. Using a small chisel, collapse the seal rim while exercising extreme care not to gouge the base metal. Install the new seal with the steep side of the lip toward the engine. Normally, seals install with the marked side visible. Seal depth is a matter of factory specification, but the seal can be displaced a tiny fraction of an inch if the old seal has grooved the crankshaft. But do not bury seals on four-cycle engines so deeply that the oil drain hole is blocked. Figure 7-41 illustrates a seal driver, which should be precisely sized to seal OD. A soft wooden block can be used as a driver when the seal mounts flush with its retainer.

Oil circuits

Oil lubricates seals and cools internal parts. Side-valve engines depend on the splash generated by the dipper on the rod cap. Because of the remoteness of the valve gear, OHV engines must have an oil pump, although small overhead

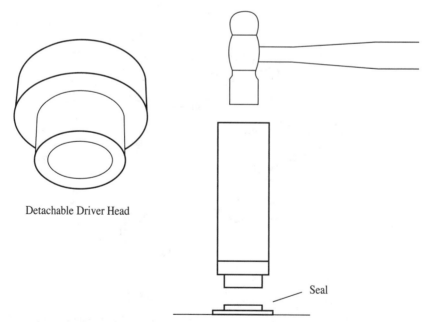

Detachable Driver Head

Seal

FIG. 7-41   *The most useful drivers have detachable heads sized for metric and U.S. seals.*

cam Hondas get around this requirement by using the cam drive belt as a conveyor. An oil pump is also necessary when an oil filter is included in the package. The more sophisticated designs pump oil to the main and crank-pin bearings by way of drilled passages in the crankshaft.

A quick check for oil pressure can be made by verifying that the overhead valves receive lubrication. The underside of the valve cover should be sopping wet with oil. A better check is to connect an oil-pressure gauge at the port for the oil-pressure sensor. Most pumps generate 25 to 35 psi at full rated engine speed. When in doubt, replace the pump.

The sludge and metal chips left over from manufacturing that collect in oil passages do not respond to ordinary cleaning techniques. It is necessary to scour these passages with pipe cleaners or the small wire brushes sold at gun shops. A reground crankshaft should have its internal oil passages thoroughly cleaned, an operation that requires removing the pressed-in plugs that seal the crank-pin drillings.

And finally, be very careful of "bargain" oil filters, many of which are counterfeits disguised as OEM parts.

Rewind starters

Rewind, or recoil, starters come in a variety of shapes and sizes (Fig. 7-42). But all have a:

- Pressed-steel or plastic housing that locates the starter concentric with the flywheel
- Recoil spring, with one end anchored to the housing and the other to the pulley
- Pulley, or sheave, sized to provide leverage and riding on a bushing in the housing
- Nylon cord
- Clutch assembly that engages the flywheel hub during starting and disengages once the engine starts.

Starter troubleshooting

Rewind starters appear to be designed to last for the life of the engine and no longer. A hard-starting engine upsets the calculation.

Broken rope This is the most common failure and one that often can be blamed on the operator who pulls too hard or at a steep angle. A worn rope bushing—the guide at the point where the rope exits the housing—contributes to the problem.

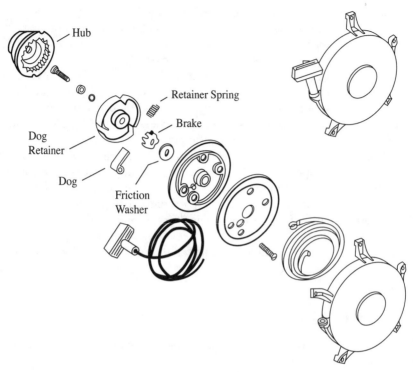

FIG. 7-42 *Eaton and Eaton knock-offs are among the most frequently encountered starters, used on Kohler and other American engines. The heavy-duty version with a two-piece pulley and engagement three dogs is shown here. The star-shaped brake should be replaced at the first sign of slippage.*

Failure of the rope to retract   A rope that hangs useless, its full length extended out of the housing, means that the recoil spring has broken or come adrift from its anchors. Partial retraction usually can be traced to a weak recoil spring.

Hard pulling   The most likely cause of rope drag is misalignment between the starter and flywheel. If repositioning the starter does not help, suspect a dry pulley bushing.

Noise when the engine runs   A squeal or rattle means a misaligned starter assembly. Loosen the hold-down capscrews, and center the starter over the flywheel hub.

Preload release and restoration

The first step in disassembly is to release the spring preload that causes the rope to retract fully. Any rewind starter can be disarmed by removing the rope handle and allowing the pulley to slowly unwind while braking the pulley with your thumb. If the pulley has a notch, preload can be dissipated with the handle attached (Fig. 7-43).

The final step in replacing the rope and other service operations is to restore preload:

1. Remove the rope handle.
2. Anchor one end of the rope on the pulley. How this is done varies, but most ropes anchor with a square knot.
3. Wind the rope completely over the pulley to produce the correct direction of flywheel rotation when the rope is unwound.
4. Wind the pulley against the direction of engine rotation. If the preload specification is unknown, wind until coil bind and slowly release the pulley, braking it with your thumbs, for one or two revolutions.
5. While holding the pulley stationary, thread the rope through the housing bushing, and attach the handle.

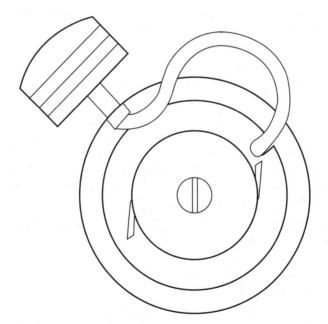

FIG. 7-43 *Preload can be released by disengaging several turns of rope from the pulley. Some Briggs & Stratton starters have enough clearance between the pulley and housing for this to be done. Other manufacturers notch the pulley as shown.*

6. Fully extend the rope to verify that it retracts and that the spring does not bind.

Note that a new rope should be the same diameter and length as the original.

Another approach can be used when the rope anchors to the engine side of the pulley:

1. Assemble the pulley and recoil spring, being careful to engage the free end of the spring with the slot in the pulley.
2. Rotate the pulley in the opposite direction of engine rotation until the recoil spring binds.
3. Slowly release spring tension for one or two pulley revolutions.
4. Block the pulley to hold spring tension. Some pulleys have provision for a nail to be inserted; others are secured with Vise-Grips.
5. Attach the handle, thread the rope into the bushing, and anchor it to the pulley with a knot.
6. Release the pulley, braking it with your thumbs. The pulley will rewind, pulling the rope after it.
7. Test the starter.

Brake

Failure to engage the flywheel cup can be corrected by tightening the dog-retainer screw hard or replacing the friction spring and associated washers.

Dogs

When present, dog springs are quite vulnerable. Briggs & Stratton plastic dog retainers routinely fail.

Recoil springs

Most recoil springs come prewound in a retainer and are simply slipped into place and anchored to the starter housing. Others have a detachable housing that is removed after the spring is safely anchored. A few European starters force one to wind the spring into place.

CAUTION: Springs can get loose and flail about. Wear safety glasses and long sleeves when servicing.

A final word

And now we have come to the end of your journey. If you have any questions about gensets or comments, good or ill, about the book, please contact me through the publisher. I'm always glad to hear from readers.

A

Generator websites

Grounding

http://amasci.com/ele-edu.html
Well-written account of basic electricity.

http://bellsouthpwp.net/j/o/johngd/files/rv/inverter_generator.pdf
Inverters theory, by someone who understands the subject.

http://home.howstuffworks.com/question117.htm
GFCI operation.

http://hyperphysics.phy-astr.gsu.edu/hbase/emcon.html#emcon
Motors, with material that applies to generators.

http://ww2.electrical-design-tutor.com/generatorgrounding.html

http://nuclearpowertraining.tpub.com/h1011v3/css/h1011v3_22.htm
Good introduction AC generator theory.

http://oshaprofessor.com/Portable%20Generators%20and%20OSHA%20
Construction%20Standards%203-05.pdf
Authoritative source for grounding requirements.

http://science.howstuffworks.com/environmental/energy/power.htm
House wiring, color codes, etc.

http://science.howstuffworks.com/environmental/energy/power.htm
Grid grounding.

http://www.animations.physics.unsw.edu.au/jw/electricmotors.html
Painless approach to electrical theory, fun to learn.

199

http://www.cumminspower.com/www/literature/technicalpapers/
PT-6006-GroundingAC-2-en.pdf
Grounding, written by an engineer.

http://www.electrical-design-tutor.com/generatorgrounding.html
Grounding separately derived systems; transfer switches.

http://www.generatorsforhomeuse.us/portable-generators/
Surge wattage ratings, ground fault detection problems, etc.

http://www.motorsanddrives.com/cowern/motorterms16.html
Description of power factor.

http://www.phys.unsw.edu.au/hsc/hsc/electric_motors.html
Motor types, functions, applications.

http://www.screenlightandgrip.com/html/emailnewsletter_generators
.html#anchorBrushless%20Generators
A major source of information on inverter generators, power factor,
and related material.

http://www.windstuffnow.com/main/3_phase_basics.htm
Describes 3-phase alternators.

http://www.usbr.gov/ssle/safety/RSHS/appC.pdf
Grounding.

Generators

http://powertech.myshopify.com/collections/8-ksi-generator-parts
Kubota parts.

http://store.eurtonelectric.com/hondageneratorrewinding-3-2-1.aspx
$265 to rewind Sawafuji generator rotors, plus a wide range of genera-
tor and other electrical parts.

http://weingartz.com/find-parts-by-illustrated-diagram/?gclid=CNSt5
_KOhLQCFWrZQgodWW4A3w
Honda factory and aftermarket parts for various generators.

http://wemakeitsafer.com/Generators-Recalls
U.S. Product Safety Commission generator recalls.

http://www.wincogen.com/Winco_Downloads/
Winco portable generator owner's manual, Honda and other engines,
wiring diagrams.

http://www.arkansas-ope.com
Parts and owner's manuals for Briggs & Stratton, Honda, Coleman,
Kubota, Kohler, Generac and others.

http://www.briggsandstratton.com/us/en/support/manuals
Briggs & Stratton operator's manuals and wiring diagrams.

http://www.electricgeneratorsdirect.com/compare.php
Parts for Generac and Briggs residential generators with information
transfer switches.

http://generatorguru.com/categories/Make-%26-Model/?gclid
Parts and repair videos.

http://www.generatorjoe.net
Excellent site, with new and used generators for sale, parts and technical advice.

http://www.gillettegenerators.com/categories/2/Portable#
Gillette generator homepage.

http://www.jackssmallengines.com
Parts and manuals.

http://www.mayberrys.com
Honda generators, some technical information.

http://www.mymowerparts.com/pdf/
Generac and other engine parts and manuals.

http://www.partsfortechs.com/asapcart/generator-parts-c-14.html
Parts for Kohler, Generac, Cat, Coleman, FG Wilson and Briggs &
Stratton.

http://www.poweredgenerators.com/guardian/manuals/0F7713.pdf
Generac portable repair manual.

http://www.repairclinic.com/
Popular parts for most generators.

http://www.rverscorner.com/onan.html
Onan troubleshooting.

http://www.youtube.com/watch?v=gQD9H0dJPKY&feature=player
_detailpage#t=0s
Honda EU2600 EU 3000 owner's manual.

http://www.youtube.com/watch?v=tC_LrrZcx38
Quick and somewhat hazardous tests for AVRs, rotors, and stators.

Engines

http://coxengineering.sharepoint.com/Pages/Boreglazing.aspx
Bore glazing in diesel engines.

http://coxengineering.sharepoint.com/Pages/Enginefaults.aspx
Troubleshooting diesel engines.

http://dawinfo.com/LV-Series-Service/LV-Series-Service-Complete.pdf
Downloadable Yanmar repair manuals.

http://engines.honda.com/models/model-detail/gx160
Information on Honda engines.

http://outdoorpowerinfo.com/
Torque and valve clearance specifications, product safety recalls, belt
sizes and how-to information.

http://stores.ebay.com/Honda-Power-Equipment-Publications
/Shop-Manuals-/_i.html?_fsub=2861039018
Honda shop manuals.

http://tinytach.com/tools.php
Specialized engine tools.

http://www.auroragenerators.com/downloads/AGI6500Manual.pdf
Good introduction to portable diesel generators, includes U-Tube links
for replacing injectors and other repairs.

http://www.auroragenerators.com/generators/portable-diesel-generators
Parts and good technical support for Chinese clones.

http://www.auroragenerators.com/index.php
Parts for Yanmar clones.

http://www.continentalengines.com/hatz-diesel-parts.html
Yanmar, Kohler, Lombardini, Subaru, and Hatz parts, specifications and
parts lists.

http://www.dieselserviceandsupply.com/Articles_info.aspx
Excellent material on diesels.

http://www.ereplacementparts.com/briggs-and-stratton-0302070-10000
-watt-portable-generator-parts-c-16758_24390_25035.html
Briggs & Stratton parts, some repair information, including rotor-
bearing removal and two-stroke fuel line routing.

http://www.gardentractorpullingtips.com/valvecam.htm
Emphasis on side-valve Kohler engines, plug but much technical information of the sort that can only be purchased by experience.

http://www.soverel33.com/Boat/Engine/Ch02/Info.html
Yanmar 1GM service manual; a useful source, although the IGM is a
marine engine.

http://www.goodson.com/Catalog/index.html#p=22
Engine rebuilders' tools.

http://www.hatz-diesel.com/index.php?id=50&L=1
Hatz downloadable specifications, parts lists, service manuals.

http://www.honda-engines-eu.com/en/images/39064.pdf
Honda Europe website with much technical information.

http://www.kohlerengines.com/manuals/landing.htm
Kohler shop manual downloads.

http://www.masenorthamerica.com/Manuals/DM/DSM-Y%20Model
%20TNE.pdf
Downloadable Yanmar multi-cylinder service manual.

http://www.mfgsupply.com/smallengine/smengtools.html
Small engine tools and manuals.

http://www.perr.com/gen.html
Genset repairs and engine families.

http://www.smallenginesuppliers.com/
Much information about popular engines, including specifications,
including valve lash and torque limits for Briggs & Stratton.

http://www.soverel33.com/Boat/Engine/Ch02/Info.html
Yanmar marine engine shop manual.

http://www.thedieselstop.com/contents/getitems.php3?Breaking%20in
%20a%20Diesel%20Engin
Breaking in a diesel engine.

http://www.westmorelandequipment.com/
Honda parts and short blocks.

http://www.westmorelandequipment.com/hondaengines.aspx#specs
Honda torque specs and more.

Carburetors

http://cortapastos.50webs.com/Tillotson%20Manual%20de%20Servicio.htm
Tillotson carburetor manual.

http://wem.walbro.com/distributors/servicemanuals/
Downloadable Walbro service manuals.

http://www.aerocorsair.com/id150.htm
Diaphragm carburetors used on ultra-light aircraft, but includes useful
information for genset owners.

http://www.drystacked.com/Walbro%20Carburetor%20Theory%2027
Jun2010.pdf
Good description of Walbro diaphragm carburetor operation and
repairs.

Fuels

http://journeytoforever.org/biodiesel_stana.html
Diesel fuels, including biofuel, additives that low-sulfur fuel requires.

http://www.propanecarbs.com/methods.html
Propane carb kits, carburetors, and other hardware.

http://www.propanecarbs.com/propane.html
Fuel specs, Btu, flash point, etc.

http://www.propanecouncil.org/research-development/resource-library
/technology-fact-sheets/
Tank coatings, pressure regulators, and generator modifications for
propane.

http://www.propane-generators.com/natural-gas-chart.htm
Piping requirements for NG generators.

Safety

http://www.doityourself.com/stry/the-dangers-of-burning-copper-wiring
Dangers posed by dioxin released from overheated wiring.

http://www.ereplacementparts.com/article/6125/How_to_Replace_Fuel
_Lines_on_2Cycle_Engines.html
Safety and maintenance instructions.

http://www.michigan.gov/documents/mpsc/portablegenerator_189253
_7.pdf
Safety tips.

http://www.esfi.org/index.cfm/page/Resource-Library/pid/10272

htp://www.usbr.gov/ssle/safety/RSHS/appC.pdf

Forums

http://forums2.gardenweb.com/forums/tools/

http://ths.gardenweb.com/forums/load/wiring/msg0821144423622.html

http://www.backwoodshome.com

http://www.doityourself.com/forum/outdoor-gasoline-power-equipment
-small-engines

http://www.generatorforum.org

http://www.perr.com/phpBB3/viewforum.php?f=12
Small engine technical forum with some very knowledgeable
contributors.

http://www.powerequipmentforum.com/forum/9-generator

http://www.smokstak.com/forum

Repair information and parts sources for vintage generators and engines.

B

National Electrical Manufacturers Association (NEMA) plug and receptacle configurations

15 AMP	2 pole 2 wire		2 pole 3 wire grounding			3 pole 3 wire		3 pole 4 wire grounding		4 wire
	125V	250V	125V	250V	277V	125/250V	30 250V	125/250V	30 250V	30 120/ 208V
Receptacle	1-15R		5-15R	6-15R	7-15R		11-15R	14-15R	15-15R	18-15R
Plug	1-15P	2-15P	5-15P	6-15P	7-15P		11-15P	14-15P	15-15P	18-15P

20 AMP	2 pole 2 wire		2 pole 3 wire grounding			3 pole 3 wire		3 pole 4 wire grounding		4 wire
	125V	250V	125V	250V	277V	125/250V	30 250V	125/250V	30 250V	30 120/ 208V
Receptacle		2-20R	5-20R	6-20R	7-20R	10-20R	11-20R	14-20R	15-20R	18-20R
Plug		2-20P	5-20P	6-20P	7-20P	10-20P	11-20P	14-20P	15-20P	18-20P

30 AMP	2 pole 2 wire		2 pole 3 wire grounding			3 pole 3 wire		3 pole 4 wire grounding		4 wire
	125V	250V	125V	250V	277V	125/250V	30 250V	125/250V	30 250V	30 120/ 208V
Receptacle		2-30R	5-30R	6-30R	7-30R	10-30R	11-30R	14-30R	15-30R	18-30R
Plug		2-30P	5-30P	6-30P	7-30P	10-30P	11-30P	14-30P	15-30P	18-30P

50 AMP	2 pole 3 wire grounding			3 pole 3 wire		3 pole 4 wire grounding		4 wire
50 AMP Receptacle	5-50R	6-50R	7-50R	10-50R	11-50R	14-50R	15-50R	18-50R
Plug	5-50P	6-50P	7-50P	10-50P	11-50P	14-50P	15-50P	18-50P

60 AMP	3 pole 4 wire grounding		4 wire
	125/250V	30 250V	30 120/ 208V
Receptacle	14-60R	15-60R	18-60R
Plug	14-60P	15-60P	18-60P

Source: Lawrence Berkeley National Laboratory

Index